T0186827

*Lillian R. Brazin, MS, AHIP*

# Internet Guide to Medical Diets and Nutrition

"Lillian Brazin has created a reputable and responsible resource of nutrition-related subjects available on the Internet for both the health professional and consumer. This informative book guides you on a journey through cyberspace, offering a balanced analysis of available nutrition sites and tips on evaluating the quality of information provided. The *Internet Guide to Medical Diets and Nutrition* provides a comprehensive overview of nutrition Web sites that is easy-to-use, well-organized, and separates fact from fiction. It looks at sites that provide nutrition tools, i.e., BMI calculators, factual nutrition information, nutrition information for diets reflecting religious or philosophical beliefs and lifestyles, as well as many sources for recipes. Concise descriptions are included for every site. As a registered dietitian, I found this guide offers a wealth of nutrition-related information, including sites that can enhance professional presentations to sites providing recipes for the health-conscious individual. Hard-to-find information is provided for many medical conditions from diabetes to seizure disorders that is both factual and practical.

As I read, I found myself looking up Web sites I never knew existed and finding information to improve my professional presentations as well as facts and statistics on obesity and nutrition-related diseases. Individuals living with celiac disease will find a very comprehensive list of resources complete with recipes and sources for purchasing the designated foods. Information in this book also includes popular weight-loss diets, surgical intervention for weight loss, weight-loss spas, and residential diet programs. From now on, the *Internet Guide to Medical Diets and Nutrition* will be next to my computer as a valuable resource for my work as a registered dietitian and certified diabetes educator."

**Nadine Uplinger, MS, MHA, RD, CDE, LDN**
*Director,*
*Gutman Diabetes Institute,*
*Albert Einstein Healthcare Network*

# Internet Guide
# to Medical Diets
# and Nutrition

*THE HAWORTH INFORMATION PRESS*®
Haworth Internet Medical Guides
M. Sandra Wood, MLS
Editor

# Internet Guide to Medical Diets and Nutrition

Lillian R. Brazin, MS, AHIP

The Haworth Information Press®
An Imprint of The Haworth Press
New York • London • Oxford

For more information on this book or to order, visit
http://www.haworthpress.com/store/product.asp?sku=5852

or call 1-800-HAWORTH (800-429-6784) in the United States and Canada
or (607) 722-5857 outside the United States and Canada

or contact orders@HaworthPress.com

Published by

The Haworth Information Press®, an imprint of The Haworth Press, Inc., 10 Alice Street, Binghamton, NY 13904-1580

PUBLISHER'S NOTES
The development, preparation, and publication of this work has been undertaken with great care. However, the Publisher, employees, editors, and agents of The Haworth Press are not responsible for any errors contained herein or for consequences that may ensue from use of materials or information contained in this work. The Haworth Press is committed to the dissemination of ideas and information according to the highest standards of intellectual freedom and the free exchange of ideas. Statements made and opinions expressed in this publication do not necessarily reflect the views of the Publisher, Directors, management, or staff of The Haworth Press, Inc., or an endorsement by them.

This book has been published solely for educational purposes and is not intended to substitute for the medical advice of a treating physician. Medicine is an ever-changing science. As new research and clinical experience broaden our knowledge, changes in treatment may be required. While many potential treatment options are made herein, some or all of the options may not be applicable to a particular individual. Therefore, the author, editor, and publisher do not accept responsibility in the event of negative consequences incurred as a result of the information presented in this book. We do not claim that this information is necessarily accurate by the rigid scientific and regulatory standards applied for medical treatment. No warranty, express or implied, is furnished with respect to the material contained in this book. The reader is urged to consult with his/her personal physician with respect to the treatment of any medical condition.

Due to the ever-changing nature of the Internet, Web site names and addresses, though verified to the best of the publisher's ability, should not be accepted as accurate without independent verification.

Cover design by Jennifer M. Gaska.

**Library of Congress Cataloging-in-Publication Data**

Brazin, Lillian R.
    Internet guide to medical diets and nutrition / Lillian R. Brazin.
        p. cm.
    Includes bibliographical references and index.
    ISBN-13: 978-0-7890-2358-2 (hc. : alk. paper)
    ISBN-10: 0-7890-2358-X (hc. : alk. paper)
    ISBN-13: 978-0-7890-2359-9 (pbk. : alk. paper)
    ISBN-10: 0-7890-2359-8 (pbk. : alk. paper)
    1. Nutrition—Computer network resources—Directories. 2. Dietetics—Computer network resources. 3. Diet therapy—Computer network resources. 4. Reducing diets—Computer network resources. 5. Weight loss—Computer network resources. I. Title.
RM217.B75 2006
615.8'54'0285—dc22
                                                                                    2006007138

To clinical dietitians, who take good nutrition seriously

# ABOUT THE AUTHOR

**Lillian R. Brazin, MS, AHIP,** is Director of Library Services at the Albert Einstein Healthcare Network in Philadelphia, Pennsylvania. She has more than 30 years experience as a medical reference librarian, including the areas of online services, user instruction, and management of academic health sciences libraries, and 10 years experience as an evening and weekend reference librarian for the Free Library of Philadelphia. Ms. Brazin has developed and taught workshops on Internet resources and is the author of *The Guide to Complementary and Alternative Medicine on the Internet* (Haworth).

# CONTENTS

# Preface

My interest in diets began when I was a child. My father was a type 1 (insulin-dependent) diabetic at a time when all medical science could offer to stave off the terrible and wide-reaching complications of this disease were painful insulin injections, frequent physician visits for blood and urine testing, and strict adherence to a bland, sugar-free diet based on measured food exchanges. When complications set in (as they eventually did for most diabetics), patients were treated with medications, such as nitroglycerin (for chest pain), or radical procedures, such as tooth extraction (for extensive gum disease) or amputation (for gangrene of the toes, feet, or legs). My father lived to his early fifties (he died of leukemia) in relatively good health because my mother prepared meals that followed the portion-controlled mix of vegetables, fruits, fats, starches, and protein from food groups in the food pyramid (the version published in the mid-twentieth century). Dad had a family doctor that specialized in treating patients with diabetes mellitus. My father was mature enough to realize he had to strictly adhere to the diet. There was a Sunday ritual at our house in which he would go to the local bakery and purchase rich confections (eclairs, sweet pastries filled with jams and poppy seeds, and exotic treats such as halvah and marzipan). He enjoyed watching my sister and me devour these lush treats!

I have colleagues who have celiac disease and must follow gluten-free diets. Many friends have osteoporosis, so they try to eat calcium-rich foods, in addition to taking medications such as Fosamax and performing weight-bearing exercises (walking, jogging, lifting weights, and doing low-impact aerobics). A former colleague, who survived a heart attack while in her early forties, is a "graduate" of the Pritikin Longevity Center, and another follows Dr. Dean Ornish's diet plan after

*Internet Guide to Medical Diets and Nutrition*
© 2006 by The Haworth Press, Inc. All rights reserved.
doi:10.1300/5852_a

undergoing a coronary artery bypass procedure. Many acquaintances follow vegetarian diets for religious and cultural reasons. Some colleagues observe kosher dietary rules.

In the past, I have used liquid meal replacements to drop a few pounds. I am a "lifetime member" of Weight Watchers, having reached my goal weight twenty years ago. I have fallen off the wagon several times, but Weight Watchers welcomes me back every time. Last year, many members of our hospital staff, including some physicians, nurses, and dietitians, followed a low-carbohydrate diet. My "Satkins" regimen (I unofficially blended the Atkins Diet with the South Beach Diet) has been the easiest diet I've ever followed. Of course, I hate to cook, and I do not mind eating ricotta cheese, salads, chicken, shellfish, soy chips, mozzarella string cheese, and soy burgers nearly every day! Until last year, I had not purchased a real egg in years (I kept packets of soy sauce in the egg keeper of my refrigerator). The first two weeks of the "induction" phase I went a little loopy over microwave bacon. Do you think eating two pounds a week is excessive? I freaked out when my cat grabbed a piece of bacon I was preparing to eat.

Who knew that cats liked bacon? I pulled the piece right out of his mouth and gobbled it up (you should never mess with a woman starting a diet).

On a more serious vein, the ketogenic (high-fat) diet is reported to have helped some children (at least temporarily) who suffer from seizures that cannot be controlled by medication. Children and adults who suffer frequent migraine headaches may benefit from diets that restrict certain foods. Some of the migraine trigger foods are aged cheeses, chocolate, red wine, and champagne.

The Internet provides a wealth of resources to make it easy for people to learn about various diets, to chat with those who have the same health problems, religious beliefs, or love of a particular cuisine. The Web sites with recipes and ingredient lists help families and individuals prepare varied, tasty, and nutritious meals following specific diet regimens. This book includes sites that teach you how to read a food label and calculators to assess your body mass index (BMI) to determine whether you are obese or only overweight. There are thousands of books on diets and recipes, but the medium of the Internet can provide truly up-to-date information and support at any time in the comfortable and private setting of your home. Wishing you good eating!

# Chapter 1

# Introduction

Do you eat to live—or live to eat? Are you digging your grave one forkful at a time? Just as gasoline fuels your car, food nourishes your body. Poor-quality food promotes a sluggish "engine." Nutritional deficiencies could cause thinning hair, blotchy skin, lack of energy, depressed mood, and more. For some people, the wrong diet can cause or exacerbate serious medical conditions such as diabetes, heart disease, allergies, and celiac disease.

One-third of adult Americans are overweight (BMIs of 25-29) and another third are obese (BMIs of 30 and greater). Annually, there are over 300,000 obesity-related deaths in this country. In the United States, obesity-related illness leads to annual medical costs of $90 billion.[1] Many Americans are on weight-loss diets. Surgical procedures for weight reduction (bariatric surgery) are performed in many hospitals. Common questions from people considering this option are these:

> Is surgery a quick solution to the problem of being overweight?
> How do I find an expert bariatric surgeon?
> What are the side effects and risks, both immediate and long-term, of such procedures?
> How successful is the surgery?
> What is the difference between obesity and overweight?
> How do I calculate my body mass index (BMI), the ratio used to determine whether a person is obese or overweight?

You may have been advised by your health care provider to follow a particular food plan or eliminate certain foods from your diet. A family

*Internet Guide to Medical Diets and Nutrition*
© 2006 by The Haworth Press, Inc. All rights reserved.
doi:10.1300/5852_01

may follow a diet for religious reasons but also need to incorporate certain food restrictions because of health problems (e.g., one family member may adhere to a vegetarian lifestyle, while another is on a low-carbohydrate regimen built around fish, fowl, and beef).

How do you locate up-to-date information?
How do you find tasty, varied recipes to prepare that follow the guidelines and restrictions of your diet?
How can the primary meal preparer plan interesting, attractive, and nutritional meals that satisfy everyone?

You could spend a lot of money purchasing a shelf full of cookbooks and nutrition advice books. A small reference library should be in every home, but after a year or two some of the material will be obsolete. Web sites for major diet programs are kept current, reflecting new knowledge and new features, such as recipes, Q&A with dietitians, and success stories.

Many of us like to interact with others following the same diet or trying to lose weight. One of the best features of the Weight Watchers program is the group meeting, but not everyone has the time to attend regular meetings. Some of us are uncomfortable attending meetings or having (somewhat) public "weigh-ins." The Weight Watchers Web site has a fee-based online program, with discounts given to traditional meeting attendees. Other diet programs feature free online forums.

I wanted to write a book that would bring together the best Web sites covering these nutrition topics. At the same time I wanted to help the beginning Internet user learn how to find and evaluate nutrition information in cyberspace. What are the criteria that serve as guidelines when sifting through the information on the Web? This book will teach the following:

- How to determine whether the information you find is correct
- How to locate Web sites recommended to you
- Where to begin to research a particular diet or health problem affected by diet
- How to use criteria to evaluate a Web site
- When to use search engines

- What is the significance of "domains" in Web site addresses
- How to observe proper etiquette while participating in Internet discussion groups

In most instances, commercial sites were excluded if they did not include, free of charge, basic information about the diet program or diet philosophy. Some sites, although maintained by board-certified physicians, exist, primarily, to sell the physicians' books. Some of the books or diet philosophies are recommended by physicians with expertise outside of fields related to nutrition or to the diseases supposedly remedied by the diet regimens. On one site, a plastic surgeon and an orthopedic surgeon write glowingly about a diet guru's books and diet plan. At this same Web site, little information is disclosed about the ground rules of the diet. A reader would have to enroll (for a fee) in the diet program or purchase the author's books. Follow your instincts. If a diet regimen sounds bizarre to you, do not even think about following it. Some diets are so flaky or dangerous or promoted by people with suspicious credentials (the proponents almost "quack"), that you know they are not healthy. Every city has expert physicians and certified dietitians. Medical reporters are available at every television station and newspaper. Heed their advice.

## NOTE

1. Vastag, B. (2004). Obesity now on everyone's plate. *Journal of the American Medical Association* 291:1186-1188.

# Chapter 2

# Hunting for Information

## *SEARCH ENGINES*

Search engines allow you to type a word, phrase, or name and retrieve Web sites and discussion group responses using these words, phrases, or names. Depending on the search engine, the "hits" that come up may or may not be relevant to your needs. In the context of this book, readers seek quality, not quantity. In most instances, they should not need to go on cyber-fishing expeditions to locate useful Web sites. Use search engines to zero in on the correct URL (Web site address) of a recommended site or a resource mentioned in the newspaper or on television. For this reason, there will be no instruction in how to construct elaborate search strategies using search engines. Just keep it simple. In most instances, type a word or phrase in the white "search" block and click the "enter" or "go" button. Some of the search engines have topical directories that help narrow the search. Begin with the preselected groups of Web sites in the directories.

**About.com**
**<http://www.about.com>**

About.com started in 1997 as The Mining Company. In 2005 the New York Times Company acquired the site. About.com features two dozen general guides written by nearly 500 outside experts who have been selected and trained to manage or contribute text to the content channels. The quality of the guides varies, so try to back up the material with information from government and medical organization Web sites.

*Internet Guide to Medical Diets and Nutrition*
© 2006 by The Haworth Press, Inc. All rights reserved.
doi:10.1300/5852_02

## Dogpile
## <http://www.dogpile.com>

Dogpile ("all the best search engines piled into one"), is a meta-search engine: it searches several resources simultaneously (Ask Jeeves, Yahoo!, Google, etc.). Dogpile was created in 1996 and is currently owned and operated by InfoSpace, Inc., a wireless and Internet solutions company marketing its services to businesses. Search by specific keywords and click the "fetch" button. You can search specific illnesses and conditions. Be aware that Dogpile is a commercial entity, accepting advertisements. Some "featured listings" are commercial sites that sell products (see Figure 2.1).

## Google
## <http://www.google.com>

Google, started in 1998 by former Stanford University graduate students Lawrence Page and Sergey Brin, has become a favorite search engine of medical librarians and physicians. The phrase "to Google someone" (conduct a Google search for information about a person) has crept into our lexicon. Google ranks Web sites by how frequently other sites link to them, and integrates the contents of several smaller search engines. Use Google Groups to locate Usenet discussion groups (Internet forums).

## Health On the Net (HON) Foundation
## <http://www.hon.ch/MedHunt>

HON is a nonprofit medical information gateway produced by the Swiss organization Health On the Net Foundation. This respected site began in 1995 during a meeting during which international telemedicine experts convened to discuss the use of the Internet in health care. Attending that meeting were officials from the National Library of Medicine and Dr. Michael DeBakey, MD, the noted heart surgeon. The attendees voted to "create a permanent body that would . . . promote the effective and reliable use of the new technologies for telemedicine in healthcare around the world." HON now works with the University Hospitals of Geneva and the Swiss Institute of Bioinformatics. The two

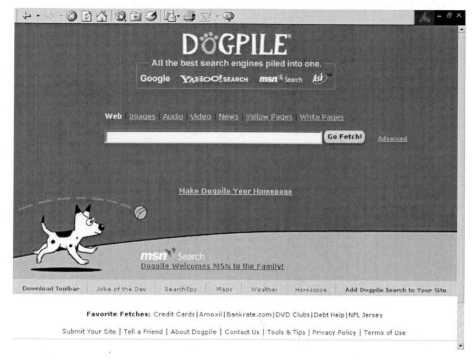

FIGURE 2.1. Dogpile
<http://www.dogpile.com>

search engines at this Web site include MedHunt and HONselect. It is considered very prestigious for a medical Web site to bear the HONcode seal of approval. The quality filtering by the HON Foundation is without compare!

## Ixquick Metasearch
## <http://www.ixquick.com>

Begun in 1998, Ixquick dubs itself the world's most powerful meta-search engine. It searches fourteen search engines at once. Ixquick highlights sponsored (paid) sites by listing them first. Large ads appear on the right side of the page.

## Mamma
## <http://www.mamma.com>

Mamma, created in 1996 as a student's master's thesis, is now owned by Intasys Corporation. The site calls itself the "mother of all search engines." Mamma is a metasearch engine, because it sifts simultaneously through several search engines. This is the search engine to use if you are looking for new, unusual, rare, or obscure topics. Click on the "Search" button after typing in a word, name, or phrase. In one respect Mamma is similar to Google: advertised Web sites are featured ("Sponsored Links").

## MetaCrawler
## <http://www.metacrawler.com>

MetaCrawler simultaneously searches several of the top search engines, including Google, Yahoo!, Ask Jeeves, and Overture. Now owned and operated by InfoSpace, Inc., MetaCrawler was originally developed in 1994 by a graduate student and a professor at the University of Washington. MetaCrawler's motto is "get better results, easier."

## Yahoo!
## <http://www.yahoo.com>

Yahoo! ("Yet Another Hierarchical Officious Oracle") was created in 1994. Its developers, two engineering students at Stanford University, originally called the search engine "Jerry's Guide to the World Wide Web." Featured Web sites are often ones suggested by users.

### *ARE YOU THE MASTER OF YOUR DOMAIN?*

The two- or three-letter tag at the end of a Web site address (com, edu, gov, net, org) indicates the origin of the site. "Com" or "net" or "biz" are usually businesses, "edu" is an educational institution such as a school or hospital, "mil" is the military, a "gov" domain tag indicates a government (local, state, or national) Web site, and "org" is an organization. Sometimes an abbreviation follows the domain. This indicates a country or foreign language ("fr" is France, "esp" is Spain or Spanish language, "ca" is Canada, and "uk" is the United Kingdom). Web sites in the "gov" or "edu" domains have the most credibility, so when in doubt, choose one of these sites. Warning: the domain designation should be noted when purchasing products over the Internet. Never respond to unsolicited e-mail messages directing you to vaguely familiar Web sites. Sometimes fraudulent sites have nearly the same URL (Internet address) as reputable sites, but the domains or country designations are different.

## *TOO GOOD TO BE TRUE:*
## *HOW TO PICK THE BEST WEB SITES*

Have you been playing cyber roulette? Would you take nutrition or medical advice from a teenager? No, you say! Well, unless you keep some guidelines in mind when searching the Internet, you might find yourself reading through an attractive Web site that contains biased, incorrect, or even dangerously outdated information. With modern, inexpensive software, anyone can create a professional-looking Web site that seems to be authoritative. How do you protect yourself? The following characteristics should be evident on good sites:

- *The identity of the Web site creator:* Is the person or group responsible for the content readily identified? Can you contact them by e-mail, fax, mail, or telephone? What are the credentials of the creator (organizational affiliation, education/training, experience, and publications)?
- *Currency:* When was the site created? When was the site last updated? Are the links to other Web sites still valid (do they work)? Is the site frequently "down" for maintenance?
- *Seal of approval:* Has the federal government or a national health organization recommended the site? Does it bear the "HONcode" logo? (This seal of approval is awarded to sites that meet the HONcode of Conduct.) *See* **Health on the Net Foundation (HON) (http://www.hon.ch)** (in this chapter). HON, created in 1995, is an international Swiss organization whose mission is to guide Internet searchers to reliable and useful online health and

medical information. It also sets ethical standards for those who develop health and medical Web sites. HON categorizes its selected Web sites as educational, individual, or commercial.

Another health Web site accrediting organization is URAC (http://www.urac.org/). "Promoting quality health care" is the motto of URAC. This nonprofit charitable organization, also known as American Accreditation Healthcare Commission, was founded in 1990. Its goal is to establish standards for the health care industry. URAC's mission is "to promote continuous improvement in the quality and efficiency of health care delivery by achieving a common understanding of excellence . . . through the establishment of standards . . . and a process of accreditation." The consumer section of the URAC site lists and links to (as of February 2006) approximately 450 sites which have received URAC accreditation. Under URAC's Health Web Site Accreditation Program, sites are evaluated for disclosure of financial backing and sponsorship, privacy and security, and quality and oversight standards.

- *Credibility and validity:* For guides to determining a health Web site's credibility and to view fraud warnings, visit Quackwatch (http://www.quackwatch.com). Another site that exposes health fraud is the National Council Against Health Fraud (http://www.ncahf.org). NCAHF is an especially good source for breaking news on health and medical fraud. Health professionals write the articles.
- *Bias:* Is the site selling books, vitamins, devices? Is it sponsored or produced by a drug company? Does the site advertise a practitioner's clinic?
- *Purpose:* Is the purpose or mission clearly indicated? Is it meant to educate, inform, support, sell a product, or attract customers or patients?
- *Audience:* Who is the intended audience: health professionals or consumers?
- *Attractiveness of the site:* Is the Web site easy to navigate and visually appealing?
- *Origin:* What is the Web site domain? "gov" and "edu" have the most credibility. Do you know the country of origin?

- *Content:* Does it provide the information you need to know and enough of what you need? Is the information accurate and current? What is the source of the information: opinion, observation, research, journal articles, or books? How does the site content compare with similar resources?

# Chapter 3

# The Internet
# or the World Wide Web:
# What's the Difference?

Internet and World Wide Web (or simply, the Web) are often used interchangeably. In fact, the Web is only a portion of the Internet. Some people with computers spend most of their time exchanging opinions in chat rooms or on newsgroups and mailing lists. They seldom look at Web sites.

## DISCUSSION GROUPS:
## CHAT ROOMS, MAILING LISTS,
## AND NEWSGROUPS

To read detailed advice on how to distinguish a mailing list (moderated or unmoderated) from a newsgroup, Usenet group, chat room, forum, bulletin board, or online support group, visit Google Groups—Basics of Usenet (http://groups.google.com/googlegroups/basics.html) or the explanations and links to other resources on Mossresource net (http://www.mossresourcenet.org).

Generally, you "subscribe" or register to send and receive messages on a mailing list, online support group, or forum. Sometimes the "list owner" or moderator sets ground rules for posting messages and participating in discussions. Generally, there are specified categories of ap-

*Internet Guide to Medical Diets and Nutrition*
© 2006 by The Haworth Press, Inc. All rights reserved.
doi:10.1300/5852_03

proved discussion topics. Some groups are "chatty" or "high traffic" and generate dozens of messages each day.

With most chat rooms, bulletin boards, and Usenet groups, there is no need to subscribe. You simply access the group's Internet site and scan current and archived (old) messages and decide whether you want to join in the message threads.

## Google Groups
## <http://google.com>

Select the section "Groups," and then type a word or phrase. Google Groups lists Usenet (newsgroups) discussion forums. The site has an excellent FAQs section that answers a beginner's questions about searching and using newsgroups.

## OneList.com

See **Yahoo! Groups** (in this chapter).

## SupportPath.com
## <http://www.supportpath.com>

This Web site provides links to support-related bulletin boards, chat rooms, local and national support organizations, meetings, and information on dozens of health problems. Formerly called Support-Group.com, its philosophy is that participation in Internet discussion groups offers the opportunity to share experiences (and not feel alone), vent emotions, and experience a feeling of hope. Participants often share personal stories.

## Yahoo! Groups
## <http://groups.yahoo.com>

Formerly OneList.com, this site is part of Yahoo.com. Here you can create your own discussion group. The service is free and there are easy instructions. Registration is required. Use Yahoo! Groups to search for online groups.

## SOME ADVICE FROM CYBER-AUNT: IS YOUR LIFE AN OPEN BOOK (OR INTERNET SITE)?

*Warning:* Group members may save Messages (archived) for public access or member access, so do not include confidential or personal information in your postings! See the next section for additional advice about participating in online discussion groups.

*Some words of caution:* It is difficult to verify the credentials (background, authority, honesty, or knowledge) of participants in online discussions. Never act on the advice offered by participants until you have checked with your nutritionist or health care provider (even if the person in the chat room says he or she is a physician or authority). Please be careful!

## PRACTICE GOOD NETIQUETTE

Because chat groups are so friendly and informal, it is tempting to act on impulse and dash off a message, forgetting basic rules of etiquette, grammar, and spelling. Also, some features are unique to the Internet. Here are some words of advice:

- Do not reveal personal or confidential information when posting a message to a group (whether a mailing list, bulletin board, chat room, or online support group). Messages may be stored (archived) indefinitely or made available to the public.
- Do not forward messages from other people without first obtaining their permission.
- Do not type in ALL CAPITAL LETTERS! This is annoying and is known as "shouting."
- Do not send replies in a fit of anger.
- Do not insult ("flame") someone over the Internet.
- Do not stalk or harass others in cyberspace.
- Carefully consider the impression your message leaves. Could your words be misunderstood as insulting or insensitive?

- Do not clog up mailboxes by posting "me too" messages. As a rule, do not reply or post a message unless you have a new or unique opinion to add.
- Reply to individuals, rather than to the entire group, when appropriate.
- There is no shame in stating that you are a "newbie" (newcomer) to discussion lists.
- Fill in the subject line when sending messages. Blank subject lines, especially on chatty (high traffic) discussion groups, waste members' time. Some moderated discussion groups require participants to use agreed-upon subject categories, so participants can delete messages which do not interest them.
- Do not be afraid to "lurk" for a while (read, but not post messages) when you first join a discussion group. It takes a while to get a feel for the "culture" of a group.
- Do not overdo emoticons (the punctuation marks used to express emotion in e-mail) or acronyms (abbreviations in which each letter stands for initials of each word).
- Examples of emoticons and acronyms follow:

  | : ( | sad |
  | : ) | happy |
  | lol | laughing out loud |
  | rotfl | rolling on the floor laughing |
  | rtfm | read the _____ manual! |
  | btw | by the way |
  | THX | thanks |
  | %-) | confused |
  | FAQ | frequently asked question |

- For additional emoticons and abbreviations, see Emoticons and Smilies Page (http://www.muller-godschalk.com/emoticon.html).
- Follow the rules on the job. Most companies have explicit guidelines about appropriate use of the Internet and e-mail. Some companies monitor employees' messages and Internet activities. When in doubt, use a home computer. It would be a shame to lose your job over Internet or e-mail policies.

# Chapter 4

# General Diet and Nutrition
# Web Sites

Be plain in dress, and sober in your diet.

Mary Wortley, Lady Montagu
(1689-1762)

No matter which diet you follow, whatever your physical status, healthy or ailing, heavy or thin, dietitians and health care providers suggest you educate yourself in the basics of good nutrition. The general Web sites in this chapter are all excellent sources of information. Most of these resources provide statistics, calculators, lists, and links to additional sites.

## American Academy of Pediatrics
## <http://www.aap.org>

This Web site contains a wealth of information on children's nutrition and obesity. Select "Children's Health Topics," browse topics in alphabetical directories, and choose "Nutrition." Readers can purchase *Guide to Your Child's Nutrition* and *Pediatric Nutrition Handbook,* but these are somewhat expensive. Pamphlets are for sale, in bulk, to pediatricians, who may offer single copies to patients at no charge: *Feeding Kids Right Isn't Always Easy* and *Starting Solid Foods.* Free, downloadable materials, such as documents on breast-feeding, sources of calcium, and anemia are listed under the section "AAP Resources."

*Internet Guide to Medical Diets and Nutrition*
© 2006 by The Haworth Press, Inc. All rights reserved.
doi:10.1300/5852_04

## American Dietetic Association on the Net/EatRight
## <http://www.eatright.org>

This is a Web site to bookmark because it is so comprehensive and authoritative. You will want to refer to it over and over again. The American Dietetic Association (ADA) is the national association of food and nutrition professionals, with over 70,000 members. One mandate is to promote health, well being, and "optimal" nutrition for the public. The site includes an archive of dozens of "Tips of the Day" that cover nutrition topics of strong interest to the public (vegetarianism, BMI) as a gauge of obesity, antioxidants, juices, iron, etc.). Type in your zip code, and the ADA lists nutrition professionals in your town. The association explains the educational credentials of registered dietitians and registered dietetic technicians. The site features timely recipes, including back-to-school lunches. Also read "Nutrition Fact Sheets," "Vegetarian Eating," "Special Needs," and "Weight Management." This last section includes "Ten Red Flags That Signal Bad Nutrition Advice." There is an illustration and detailed explanation of the U.S. Department of Food and Agriculture's current food guide pyramid. The ADA site also provides a link to the government's food guide pyramid for children. The "Product Catalog" sells books and book packages (money saving collections of materials on related subjects such as pediatric nutrition and sports nutrition) (see Figure 4.1).

## American Egg Board
## <http://www.aeb.org>

This is "the incredible, edible egg" site. The American Egg Board represents egg producers (the farming groups, not the chickens) from across the United States. Members are appointed by the secretary of agriculture. The mission is to increase the demand for eggs and egg products. As with some other food-marketing Web sites, it also makes an effort to offer educational information for consumers, kids, and teachers, recipes, fun facts, food preparation safety tips, and a wealth of nutrition facts and figures in a colorful, modern site. The FAQ section alone would be worth bookmarking for cooks, teachers, and reference librarians ("What's the difference between a brown egg and a white egg?").

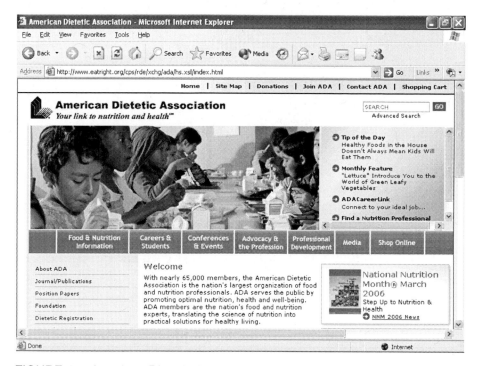

FIGURE 4.1. American Dietetic Association
<www.eatright.org>
Reprinted with permission of the American Dietetic Association.

## Ask the Dietitian
## <http://www.dietitian.com>

Joanne Larsen, MS, RD, LD, a registered and licensed dietitian, created and maintains this Web site. It contains an extensive Q&A list of medical conditions, sports, vitamin, and other nutrition- related questions that have been posted by the public. For each question, Ms. Larsen gives a thorough answer with detailed recommendations. "Top Ten Tips to Spot Nutrition Quackery" and "Ten Changes You Can Make to Lose Weight" are some of the snappy features of this nutrition site. Ms. Larsen has written books, nutrition software and databases, and appears on television and radio. Her site demonstrates her "cut to the chase" approach to nutrition advice. This is a great source for personalized, free nutrition information.

## Beech Nut Baby Food: Feeding by Age
### <http://www.beechnut.com>

Beech-Nut is a manufacturer of baby food. Its Web site is an excellent source of advice on nutrition and feeding of infants, babies, and toddlers. Select the link "Feeding by Age." Download a free pamphlet offering guidance about when a baby is ready to eat solid food. Overall, it's a warm, friendly site that offers sound information. It is also available in Spanish.

## The Blonz Guide to Nutrition, Food and Health Resources
### <http://www.blonz.com>

Dr. Edward Blonz has an MS and a PhD in nutrition from the University of California at Davis. He is a member of the Dietary Supplement Advisory Council of the U.S. Food and Drug Administration. Dr. Blonz is a certified nutrition specialist who appears on national television and writes a newspaper column and books. His Web site has won numerous awards. Dr. Blonz's site includes links to nutrition resources containing "science-reliable" evidence. According to Dr. Blonz, all else is "cyberjunk." He groups the links into categories: government; food resources and associations (everything from Taco Bell to Godiva Chocolate Web sites are linked here); nutrition, food, and fitness; search engines; academic institutions with food and nutrition programs; sustainable agriculture and gardening, and others. Quirky but authoritative!

## Calorie Control Council
### <http://www.caloriecontrol.org>

The Calorie Control Council, established in 1966, is an international, nonprofit organization representing low-fat and reduced calorie food and beverage manufacturers and suppliers. It encourages scientific research, particularly related to mutagenicity, metabolism, and carcinogenicity. The site features recipes, a free template of a daily food diary, online calorie calculators, a BMI calculator, healthy weight calculator, and general information on cutting calories and fat. Hot topics include healthy low-carb dieting and top news stories relating to diet and nutrition.

## Center for Food Safety and Applied Nutrition
## <http://vm.cfsan.fda.gov>

This is part of the U.S. Food and Drug Administration in College Park, Maryland. This is the link to the government food safety information Web site, which culls relevant information from the Centers for Disease Control, the U.S. Food and Drug Administration, the Environmental Protection Agency, and other state and federal agencies. The center's site also links to specific food safety programs and concerns involving produce imports, listeria contamination, and eggs.

## Center for Nutrition Policy and Promotion (CNPP)
## <http://www.cnpp.usda.gov>

"Improving the nutrition and well-being of Americans." This is the home of the controversial new food pyramid (MyPyramid) and "Dietary Guidelines for Americans." CNPP is part of the U.S. Department of Agriculture. Its mission is to link scientific research to the nutrition needs of American consumers and to translate the research into information and educational materials for policymakers, consumers, and health, education, and industry professionals (see Figure 4.2).

## Center for Science in the Public Interest
## <http://www.cspinet.org>

From the home page, read about hot topics in the news such as Mad Cow Disease or the Food-Allergen Labeling Bill. Select "Nutrition and Health" to access the "Health, Nutrition, and Diet" page with calculators, lists of ten foods you should or should *never* eat, ten steps to a healthy diet, campaigns concerning trans fats, sugar, ADHD and diet, and caffeine.

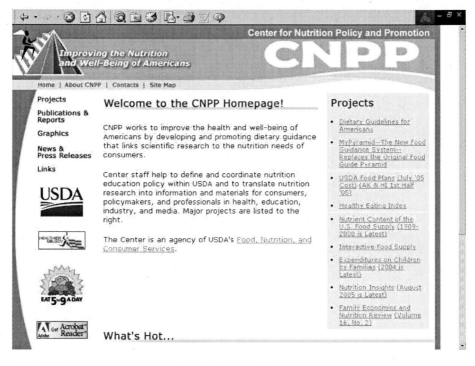

FIGURE 4.2. Center for Nutrition Policy and Promotion
<http://www.cnpp.usda.gov>

## Dole 5 A Day
## <http://www.dole5aday.com>

This is a Web site of the Dole Company. The mission of the program is to develop childhood nutrition education programs, encourage the eating of five to nine servings of fruits and vegetables per day to promote better health, and to present papers on the topic at conferences. The Web site is an especially good resource for teachers. Everything needed to design a nutrition curriculum for children is conveniently located at this site. The section "Reference Center" gives facts about the nutritional content of various fruits and vegetables. The site has charts showing the phytochemicals in fruits and vegetables, a link to the latest news about research on fruit and vegetable nutrition, and a comprehensive chart showing the nutritional values for dozens of fruits and vegetables. The "Teachers" segment has a community message board, kids' cookbook, and games.

## Egg Nutrition Center
## <http://www.enc-online.org>

The Egg Nutrition Center site is more scholarly than the American Egg Board Web site. Since it began in 1979, the center exists under a cooperative arrangement between the American Egg Board and the United Egg Producers, working to benefit the egg industry, health professionals, and consumers. The center regularly fosters communication among health professionals, consumers, producers, government agencies, and the media. A panel of independent scientists guides the center's research and education projects. The center reports the results of scientific research studies on the nutritional benefits of eggs. It publicizes information from government organizations and national medical societies. The center published the American Heart Association's new guidelines permitting consumption of up to one egg a day. The Egg Nutrition Center publishes the downloadable quarterly newsletter *Nutrition Close-Up,* which critiques and reviews the latest research on health and diet. See "A Checklist for Good Health and Nutrition," which is based on the government's new MyPyramid food guide.

## Food and Nutrition Information Center (FNIC)
## <http://www.nal.usda.gov/fnic>

This is the place to go to find the USDA MyPyramid, formerly called the Food Guide Pyramid. The pyramid was revised and renamed in January 2005. The public was invited to submit ideas and comments prior to the release of the new guide. *See* **MyPyramid.gov** (in this chapter) for more details about using the interactive MyPyramid. The Food and Nutrition Information Center (FNIC) is part of the National Agricultural Library and has been providing sound nutrition information to consumers, teachers, government employees, and health professionals since 1971. At the FNIC site are research reports, fact sheets, government reports, and educational materials. The 2005 sixth edition of *Dietary Guidelines for Americans* is available in full text on this site. The

Web site includes several related pamphlets designed to help Americans put the guidelines into practice. Note the list of downloadable food guidelines from around the world, for example, Canada, Norway, Greece, Singapore, and Great Britain. Other sections include Dietary Supplements and Herbal Information, Alternative Medicine, and a database listing the nutrient composition of over 6,600 foods. FNIC packs a lot of useful, science-based information into its Web site. Although the new pyramid has invoked much criticism, in an area rife with quackery and false claims, this government resource is outstanding for its common sense and integrity.

### Gateway to Government Nutrition Sites/Nutrition.gov
### <http://www.nutrition.gov>

This site pulls together all government nutrition resources and attempts to fulfill President George W. Bush's HealthierUS mandate by promoting physical fitness, nutrition, prevention (health screening), and avoidance of risky behaviors. The sections "What I Need to Know about Eating and Diabetes" and "The Low-Down On Osteoporosis" are particularly informative. The section "Shopping, Cooking, and Meal Planning" consolidates information on meal planning, shopping, food storage, food preservation, food labeling, ethnic cooking, and cooking methods. There is also a section on recipes, including cooking for children, heart-healthy African-American and Latino recipes, and thrifty meals.

### International Food Information Council (IFIC)
### <http://www.ific.org>

IFIC describes itself as "your nutrition and food safety resource." The site is available in Spanish, as well as English. IFIC, through the IFIC Foundation, provides science-based information to government officials, nutrition and health professionals, and journalists who educate the public about nutrition and food safety. The foundation aims to "bridge the gap" between science and communications. IFIC is supported by the food, agricultural, and beverage industries. There is a glossary of food-related terms. Through two separate sections: "Go to Food Safety Information" and "Go to Nutrition Information" the IFIC Web site provides a wealth of reliable, current reference information. Teachers, especially, will find these resources valuable.

## MyPyramid.gov
## <http://www.mypyramid.gov>

The widely touted 2005 recasting of the Food Pyramid was released with much fanfare. The online-only (at this point), interactive **MyPyramid.gov** allows individuals to personalize diet recommendations based on activity level, age, and gender. Because so many visitors tried to access the site on the first day, it crashed. Critics say its required Internet access excludes many low-income consumers, that **MyPyramid.gov** places too much emphasis on dairy foods, and that the personalization does not take into account height and weight. At one point, **MyPyramid.gov** was even part of the story line of a popular syndicated newspaper cartoon series. A separate food pyramid is available for children. See additional information under Food and Nutrition Information Council (see Figure 4.3).

## National Academy of Sciences/Food and Nutrition Board
## <http://www.iom.edu/boards.asp>

The Food and Nutrition Board is part of the Institute of Medicine, one of the National Academies of Science. The board was established in 1940, and its mission is to study issues related to the adequacy and safety of the U.S. food supply, establish guidelines for "adequate" nutrition, and render judgments on the relationship between health, nutrition, and food intake. Unlike FNIC, it does not answer questions from consumers or serve as a nutrition reference resource. It sponsors symposia and conducts research projects. Some current projects include research on how food marketing influences the diet of children, prevention of childhood obesity, and mineral requirements for physical and cognitive performance of military personnel.

FIGURE 4.3. MyPyramid.gov
<http://www.mypyramid.gov>


## National Cattlemen's Beef Association/BeefUSA.org
## <www.beef.org>


BeefUSA.org is the gateway to Web sites for the beef industry and professional cattlemen. The site clearly indicates the source of funding for its various sites. A significant portion of BeefUSA deals with the professional interests of the industry (notices of conventions, government issues affecting the industry, and cattle industry news), but there are consumer-oriented links to recipes, nutrition (http://www.beef nutrition.org), news on serious cattle diseases such as bovine spongiform encephalopathy (BSE) ("mad cow disease"), and "Zip for Tweens" (games and recipes for 'tweenagers).

## National Chicken Council (NCC). Eatchicken.com
## <http://www.eatchicken.com>

The National Chicken Council and the U.S. Poultry and Egg Association produce this Web site. Both organizations represent producers and other members of the egg and poultry industry. Both organizations follow government legislation affecting their industry and try to enlighten consumers on the nutritional value of poultry and egg consumption. The NCC site is an excellent resource for recipes and nutritional information on poultry. Every two years the National Chicken Council sponsors a national chicken cooking contest and sells an inexpensive paperback cookbook composed of the statewide-level winners' recipes, the previous national winner, and First Ladies' chicken recipes.

Eatchicken.com is a very attractive, informative site. The photographs of the recipes are tempting, information is given on the fat content of various meats, it shows a chart showing the calorie count and fat content of different meats and fish, tips on preparing chicken, and a downloadable brochure with nutrition, storage, preparation, and serving tips.

## National Dairy Council (NDC)
## <http://www.nationaldairycouncil.org>

The National Dairy Council is the nutrition marketing arm of Dairy Management, Inc., which produces the "3-A-Day Dairy" program with the American Dairy Council. NDC has been in existence since 1915. Through its nutritionists and dietitians, NDC provides science-based nutrition information to health professionals, teachers, consumers, and the media. Through its network of state and regional dairy councils, it educates schoolchildren. The Web site includes recipes and health tips. NDC's program, "Healthy Weight With Dairy," is designed around several recent research studies reporting a link between consumption of three to four servings a day of dairy products and more effective weight loss, less abdominal fat, and lower incidence of kidney stones. The site includes a BMI calculator, a link to the new federal Food Guidance System, including MyPyramid, and a database of nutritional values (calories and nutrients in foods). There is also a section on lactose intolerance. *See also* **3-A-Day.org** (in this chapter).

## National Institute of Nutrition (NIN)
### <http://www.nin.ca>

NIN is a Canadian organization bringing together nutritional science with government, industry, and consumers to produce communication programs and research projects. Its mission is to serve as the "catalyst for advancing the nutritional health of Canadians." The Web site is available in French or English. On this "plain vanilla" Web site (contains mostly text), visitors can choose from the many topics on the right side of the home page (Canada Food Guide, Dental Health, Diabetes, Dietary Fat, Vegetarian Eating, Nutrition Policy, Osteoporosis, and more). In the FAQ section are links for locating registered dietitians in Canada ("Find-a-Dietitian" and "Dial-a-Dietitian"). The site includes full-text articles from nutrition journals produced by NIN.

## National Pasta Association
### <http://www.ilovepasta.org>

This site combines consumer-friendly sections containing hundreds of recipes, an illustrated glossary of pasta shapes ("Pasta Shapes 101"), and advice on matching type of sauce with type/ shape of pasta, and nutrition discussions on folic acid, complex carbohydrates, and healthy dieting. The FAQ section is a trivia buff's dream (e.g., Thomas Jefferson brought the first macaroni machine to America. A Frenchman built the first industrial pasta company in the United States and dried his pasta on the roof of a building).

## National Turkey Federation (NTF). Eatturkey.com
### <http://www.eatturkey.com>

NTF is the national advocacy organization for the turkey industry. The motto of Eatturkey.com is "Turkey. The Perfect Protein." Besides providing nutrition information, and tips on storing, purchasing, and preparing turkey, a unique feature of this resource is a keyword searchable database of over 700 turkey recipes! Consumers can sign up to receive new recipes by e-mail. Food service professionals can observe interactive demonstrations with chefs and a virtual menu series. NTF, in collaboration with the American Culinary Federation, developed a series of modules for training food service professionals. These downloadable modules on safe food handling, serving and presentation, nutrition, and a food service recipe database are useful to schools offering certificates in culinary fields.

## Nutrition and Your Health: Dietary Guidelines for Americans
**<http://www.nal.usda.gov/fnic/dga/index.html>**

*See* **Food and Nutrition Information Center** (in this chapter).

## Nutrition Fact Sheets
**<http://www.eatright.org/public>**

A wealth of useful information is offered in this section of the American Dietetic Association's Web site. Some of the free fact sheets, downloadable in pdf format, include The Power of Potatoes, The Pasta Meal, Lycopene, Canned Food, Beef, Shopping Solutions, and African-American Health and Dairy Foods. When it comes to nutrition, dietitians are the experts, with no hidden agendas!

## Oldways: Healthy Eating Pyramids
**<http://www.oldwayspt.org>**

Oldways Preservation and Exchange Trust, the self-described "food issues think tank," is a Boston-based, nonprofit organization founded in 1990 by K. Dun Gifford, a Harvard-educated attorney who has owned and operated several restaurants. Gifford has worked with Senator Edward Kennedy and the late Senator Robert Kennedy, and he is the former national chair of the American Institute of Wine and Food. Oldways combines nutritional science with fine cuisine, educating consumers, government officials, farmers, members of the food industry, and chefs. Oldways believes that traditional foods and eating patterns are healthier than modern Western ways of eating. It offers symposia, conferences, and tours on the subject. In recent years, in collaboration with the Harvard University School of Public Health, it offered weeklong continuing education programs in Italy for physicians, dietitians, and allied health professionals who wished to study the Mediterranean diet. With input from 500 scientists, it developed five eating pyramids: Mediterranean, Asian, Latin American, and vegetarian, plus a fusion called the Eatwise Food Pyramid. To download the healthy eating pyramids, select "Wise Eating" and then "Traditional Diet Pyramids." Oldways' Eatwise program is designed to help families eat less junk food and follow healthier eating habits. The program emphasizes exercise and whole grains, while realistically recognizing that people need to accommodate eating frequent abundant or "feast" meals in their lives. Feast-type meals include parties, picnics, and restaurant meals. Oldways developed a cooking and nutrition program for schoolchildren. This site features recipes.

## Physicians Committee for Responsible Medicine (PCRM)
<http://www.pcrm.org>

PCRM is a nonprofit organization of physicians and laypeople, founded in 1985. The current president is Neal D. Barnard, MD. Barnard has published several books, and often lectures to groups of physicians. He is a dynamic speaker and makes his point in a reasonable, nonstrident way. PCRM's advisory board includes physicians, nutritionists, and scientists. The organization promotes preventive medicine, with an emphasis on healthy nutrition with a vegetarian slant. PCRM introduced the "New Four Food Groups": fruits, legumes (peas, beans, and lentils), whole grains, and vegetables. In 2005 it began a public service announcement campaign disputing the National Dairy Council and several dairy food manufacturers' linking of dairy food consumption with weight loss. In 2004 PCRM launched a vigorous print and television campaign against low-carb, high-protein diets, specifically the Atkins diet program. It maintains a registry to document adverse effects of such diets. Current initiatives include the Healthy School Lunch campaign, alternatives to animal dissection in health science education, and the Cancer Project, which studies prevention of cancer and increasing survival time for cancer patients by means of better nutrition. PCRM offers vegetarian cooking classes for cancer survivors. Download the textbook for the classes, *The Survivor's Handbook,* from the PCRM Web site. The Strong Bones campaign disputes the theory that milk is essential for strong bones ("Milk: It's Not All It's

Cracked Up To Be"). The PCRM Web site includes fact sheets on nutrition topics, including *Guide to Healthier Weight Loss,* a three-week, low-fat, vegetarian diet plan.

From the home page select "Health." Under "Preventive Medicine and Nutrition" click on "go" to see lists of downloadable fact sheets on prevention and management of many medical conditions through healthier eating (Alzheimer's disease, cancer, high cholesterol, migraine headaches, osteoporosis, menopausal symptoms, heart disease, stroke, kidney disease, arthritis, Parkinson's disease, and multiple sclerosis). To download "Vegetarian Starter Kit," select "Vegetarian Diet" after selecting "Health" from the home page. There are additional vegetarian materials here, too.

## Pork: The Other White Meat
<http://www.otherwhitemeat.com>

## Pork and Health
<http://www.porkandhealth.org>

These sites are a service of the National Pork Board. Pork: The Other White Meat is available in English or Spanish. Both resources feature recipes, instruction on safe food preparation and storage, nutrition, and education about various cuts of pork. The section "Health Professionals" links to the "Pork and Health" Web site, which offers downloadable patient education materials, including a brochure for patients with type 2 diabetes. Nutritionists contribute information on importance of protein in a healthy diet. A free e-mail newsletter, *Healthy Headlines* is also available. A separate link for teachers ("Pork4Kids") offers many classroom activities at different levels to teach about the farm-to-table process and nutrition. Also available are a food pyramid trivia game, "cyber" farm tour, "Count the Hidden Piglets," and a cartoon featuring Peggy the Pork Chop ("Are you in my food group?").

## 3-A-Day of Dairy
<http://www.3aday.org>

Dairy Management, Inc. for the American Dairy Association manages this resource. The site includes recipes, news items, and advice on nutrition. *Get 3!* is a free online newsletter providing recipes, news items, nutrition information, and advice from registered dietitians. For health professionals, the site contains digests of research studies reported in medical and nutrition journals. The 3-A-Day of Dairy campaign receives support from the American Academy of Family Physicians, the American Academy of Pediatrics, the National Medical Association, and the American Dietetic Association.

## U.S. Dept. of Agriculture National Nutrient Database (NND) from Nutrition Data Laboratory
### <http://www.ars.usda.gov/ba/bhnrc/ndl>

The databases on this site are excellent resources for learning about the nutrient content of foods. Using the Nutrient Lists database, type "calcium," for example, and see a list of the calcium content of hundreds of foods. Search the USDA National Nutrient Database for Standard Reference, and using "apricot," as an example, choose "apricots, raw" and view a detailed listing of the nutrients in one raw apricot.

## Wheat Foods Council. Grain Nutrition Information Center
### <http://www.wheatfoods.org>

This national nonprofit organization maintains the Grain Nutrition Information Center at this site. With the publication of MyPyramid, there has been renewed emphasis on consuming whole grains as part of a healthy diet. The mission of the Wheat Foods Council is to increase the public awareness of the healthful qualities of this food group. The site includes recipes.

Chapter 5

# Health Assessment Web Sites

No need for a handheld calculator, math skills, or pen and paper. These resources simplify your calculations. *See also* Chapter 4, "General Diet and Nutrition Web Sites" for additional information about calories, nutrient content of foods, and BMIs.

## *ONLINE CALCULATORS AND ASSESSMENT TOOLS*

### Basal Metabolism Calculator
### <http://www.room42.com/nutrition/basal.shtml>

Just plug in your height, weight, age, sex, and activity level and Room42's site will tell if you are overweight, obese, skinny, or just right! Cute site (tells how daughter helped dad make calculations needed to refine his diabetic nutritional requirements), but remember that the developer of the algorithm used for the calculations is not a medical professional. Wisely, the site also links to the USDA Dietary Guidelines and Height and Weight Tables.

### Calorie Control Council
### <http://www.caloriecontrol.org>

A number of calculators are available at this site: body mass index calculator, exercise calculator, two calorie calculators, weight-mainte-

*Internet Guide to Medical Diets and Nutrition*
© 2006 by The Haworth Press, Inc. All rights reserved.
doi:10.1300/5852_05

nance calculator, and healthy weight calculator. Also offers recipe section, articles on weight-loss topics, and digests of hot news related to weight issues and health.

## Fast Food Finder/Fast Food Facts
## <http://www.olen.com/food/book.html>

The Office of the Minnesota Attorney General has assembled a list of the nutrient contents of several popular fast-food items; the list includes amount of calories, fat, cholesterol, and sodium (salt).

## Lightnfit.com
## <http://www.lightnfit.com>

Light 'n Fit yogurt offers a BMI calculator and a calcium calculator on its Web site and exercise and general weight-loss tips and aids.

## Nutrition Analysis Tools and System's Energy Calculator
## <http://nat.crgq.com/energy/daily.html>

In 1996, Chris Hewes, a programmer, and Jim Painter, a registered dietitian from the Department of Food Science and Human Nutrition at the University of Illinois developed the Nutrition Analysis Tool (NAT) software. The energy calculator takes input of age, sex, height, weight, and amount of time engaging in various levels of activity during a twenty-four hour period, and then calculates the number of calories expended.

## RealAge Nutrition Health Assessment
## <http://www.realage.com>

RealAge is a consumer health media company. The Web site and the health assessment tools are sponsored by several pharmaceutical and diet program companies. Begin by answering the lengthy health and lifestyle questionnaire that gives a tally as to how closely your chronological age matches biological age. Follow the link to complete an in-depth nutrition health assessment. There is a great deal of good advice and positive encouragement at this Web site (see Figure 5.1).

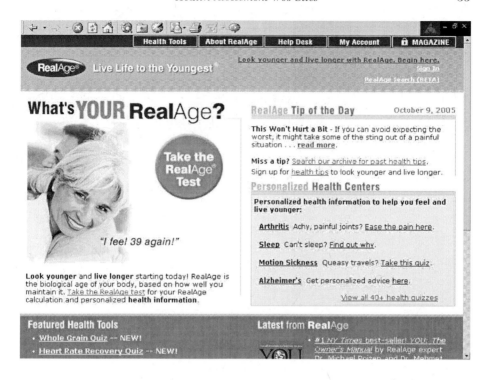

FIGURE 5.1. RealAge Nutrition Health Assessment
<http://www.realage.com>
Reprinted with permission of RealAge, Inc.

## Shape Up America! Childhood Obesity Assessment Calculator
### <http://www.shapeup.org/oap/entry.php>

Shape Up America! (SUA) is a nonprofit organization launched in 1994 by former U.S. Surgeon General C. Everett Koop to deal with the problem of overweight in America. Through a collaborative effort of experts and organizations in the fields of health, nutrition, and physical fitness, SUA promotes healthy weight and increased physical activity. The Pediatric BMI Assessment Tool is a good start to dealing with the epidemic of obesity in children.

## UNDERSTANDING FOOD LABELS

Beginning in 1990, when the Nutrition Labeling and Education Act called for overhaul of food labeling, the Food and Drug Administration and the U.S. Department of Agriculture worked to develop a more informative, easier-to-read food label that would appear on more food products and use standardized terminology. The new labels were put into use in 1994. Labeling information helps consumers calculate diet points (Weight Watchers' dieters), food exchanges (diabetics), fats (low-fat dieters, Weight Watchers' dieters, and heart-disease patients), carbohydrates (Atkins' dieters, Zone dieters, South Beach dieters, and diabetics), sodium (kidney disease patients, heart-disease patients, and hypertension patients), protein (Zone dieters, kidney-disease patients), and calcium (osteoporosis patients).

**American Diabetes Association. Reading Food Labels**
**<http://www.diabetes.org/nutrition-and-**
**    recipes/nutrition/foodlabel.jsp>**

Note the advice about serving size on this site. The Web site gives additional information important to patients with diabetes. Sugar-free foods or low-fat foods, for example, may be high in carbohydrates.

**Food and Drug Administration. Center for Food Safety**
**    and Applied Nutrition. How to Understand and Use**
**    the Nutrition Facts Label**
**<http://www.cfsan.fda.gov/~dms/foodlab.html>**

This site has an extensive tutorial on reading and understanding food labels. It also offers a video segment and supplementary downloadable documents (see Figure 5.2).

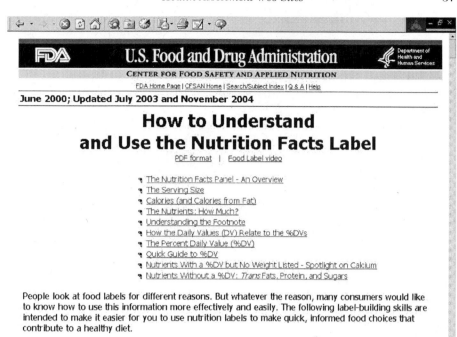

FIGURE 5.2. How to Understand and Use the Nutrition Facts Label
<http://www.cfsan.fda.gov/~dms/foodlab.html>

## MayoClinic.com. Reading Food Labels
**<http://www.mayoclinic.com>**

Select "Healthy Living" and then "Food and Nutrition Center" to read material on food labeling, including an interactive instructional session on reading food labels.

## Nemours Foundation. KidsHealth for Parents. Deciphering Food Labels
**<http://kidshealth.org/parent>**

Select the "Nutrition and Fitness" section, then "Deciphering Food Labels." Kidshealth, produced by the Nemours Foundation, is an award-winning Web site. The section on food labels describes the history of the new food-labeling system. Recipes and materials are written for children and teens, too.

## Nephrogenic Diabetes Insipidus Foundation. Label Reading Tips <http://www.ndif.org/na6.html>

This site has good advice on finding hidden sodium in foods. Note the list of ingredients, such as baking soda, brine, and MSG, which consumers might not recognize as containing sodium.

Chapter 6

# Weight Loss (Nonsurgical) Web Sites

One must eat to live and not live to eat.

Moliere

A study reported in *Archives of Pediatrics and Adolescent Medicine* in 2004 compared adolescents in thirteen European countries, Israel, and the United States.[1] The United States had the highest prevalence of overweight adolescents. The study based its findings on body mass index calculation rather than height/weight charts.

Only about one-third of American adults are of normal weight (BMIs below 25). Obese adults in the United States have 36 percent higher average annual medical expenditures compared with those of normal weight.[2] The nation spent $75.1 billion in 2003 on drugs, doctor visits, and hospitalizations related to obesity (and taxpayers footed half this bill through the government's Medicare and Medicaid medical programs).[3] In a September 2005 editorial in *The American Journal of Medicine,* Joseph S. Alpert, MD, and Pamela J. Powers, MPH, lamented, "We have a major health crisis in the United States: Americans are far too fat."[4]

Some of the most successful weight-loss programs employ teams of physicians, psychologists, and dietitians to keep their programs up to date and offer clients the Web services they expect (food product ordering, recipes, menu plans, exercise programs, one-to-one nutritional counseling, online support groups, discount coupons for products, in-person meetings, and success stories).

*Internet Guide to Medical Diets and Nutrition*
© 2006 by The Haworth Press, Inc. All rights reserved.
doi:10.1300/5852_06

A *general warning:* Although most of the programs covered in the following Web sites include a variety of food groups and encourage physical activity, everyone should have a physical examination and seek advice from a health care practitioner *before* beginning a weight-loss regimen. A few of the programs are controversial, even extreme, in that they advise the exclusion of one or more food groups or consist entirely of one food group. Be especially cautious about adopting one of these programs.

Herbal and pharmaceutical weight-loss products are not addressed in this book because of the many side effects and/or lack of scientific research on their effectiveness.

## American Obesity Association (AOA)
## <http://www.obesity.org>

AOA's Web site is a portal to research, legislation, statistics, consumer information, tax and insurance issues, and discrimination related to obesity. Print the fact sheets and read personal stories. The site is supported by law firms, pharmaceutical companies, food companies, health professional societies, and diet program companies. Note the list of medical conditions linked to obesity, including nine types of cancer (breast, colorectal, endometrium, esophagus, gallbladder, liver, pancreas, renal cell, and top of the stomach).

## American Society of Bariatric Physicians
## <http://www.asbp.org>

This organization was founded in 1950. Membership is limited to licensed physicians, and the mission is "to advance and support the physician's role in treating overweight patients." The society offers continuing medical education courses for its members. The site allows consumers to locate member physicians in local areas. The FAQ section provides information on obesity and its treatment, descriptions of weight-loss medications, statistics on prevalence of obesity in the United States, and the medical complications of obesity.

## American Society of Clinical Hypnosis (ASCH)
## <http://www.asch.net>

ASCH is one of two national organizations (*see also* **Society for Clinical and Experimental Hypnosis** in this chapter) for licensed health professionals using hypnosis.* Use this site to locate hypnotherapists in your area. Be sure to read the sections "Selecting a Qualified Hypnotherapist," "Myths About Hypnosis," and "When Will Hypnosis Be Beneficial?"

The site makes a distinction between lay hypnotists and licensed health care professionals (physicians, psychologists, dentists, nurses, and social workers) who use hypnosis as one of many treatment modalities. ASCH recommends that consumers seek hypnotherapy from licensed health care professionals.

## Atkins Nutritionals
## <http://atkins.com>

The Atkins low-carbohydrate diet has evolved over the years. The Web site is one of the best because many of the resources are free, and the site is kept current. It bridges the gap between the content contained in the Atkins diet books and the reality of going on the diet and facing questions or problems not resolved by reading the books. The success stories are varied and inspirational. The site retains the spirit of the late Dr. Robert C. Atkins, MD, featuring his biography and photographs. The company has moved on, with new books (one dealing with diabetes and the Atkins diet), and expansion of its line of food and nutritional products. Diet followers can have meals delivered to their homes in some parts of the country ("Atkins At Home"). Portions of the site are commercial, with ads for books, foods, and links to the largely fee-based eDiets Web site, but the basic sections contain helpful information for followers of the program. Atkins Nutritionals Web site features news stories, events, and recipes. Sign up to receive a free e-mail newsletter. An online support group is offered and readers can contact infor-

---

*Hypnosis is an altered state of consciousness, a deep, inner absorption with heightened concentration. Some individuals are more hypnotizable than others are.

mation representatives by e-mail or by telephoning the Atkins 800 number. The Atkins Physicians Council has developed the Atkins Lifestyle Food Guide Pyramid, which differs from the USDA MyPyramid, with protein sources (fish, beef, poultry, pork, and soy products) at the base, and whole-grain carbohydrates (barley, oats, and brown rice) at the point. Overall, it's a well-designed, colorful site with features to meet every need. The company filed for Chapter 11 bankruptcy in mid-2005, but this did not seem to affect consumers. The company plans to concentrate on its successful line of foods, including nutrition bars, cereals, sauces, and chips (see Figure 6.1).

*See also* **Carbohydrate Addict's Official Web site; South Beach Diet Online; The Zone Diet** (in this chapter).

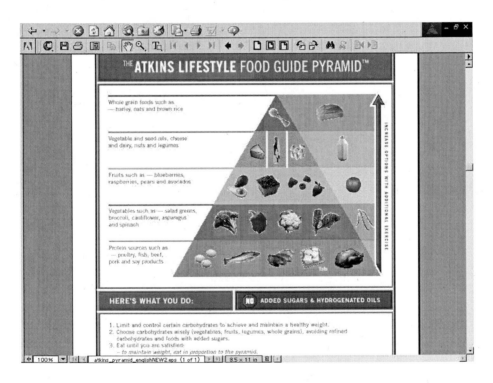

FIGURE 6.1. Atkins Lifestyle Food Pyramid Guide
<http://atkins.com>
Reprinted with permission of Atkins Nutritionals, Inc.

## BestDietForMe.com
## <http://www.bestdietforme.com>

For consumers having difficulty selecting which weight-loss diet to follow, this site can help narrow the choices. Marketdata Enterprises, a research publishing and consulting firm, created this diet-analysis tool. The service is free (advertisers subsidize the site), and the questionnaire only requires a few minutes to complete. The questionnaire covers age, weight, gender, height, budget preference, psychological eating issues, and other personal habits, attitudes, and requirements as they relate to selection of a diet program. BestDietForMe.com covers sixty popular U.S.-based diet programs. A personalized diet analysis appears on the Web site as soon as the visitor completes the survey. Depending on the responses to the questions, BestDietForMe lists many or just a few diet programs for consideration. Marketdata says a dietitian and a clinical psychologist developed the diet survey.

Even if you have already selected a weight-loss program or are currently on a diet that is satisfactory, the information on weight loss, diet fraud, dieting tips, eating disorders, and weight-loss associations available at this site will be useful. Some of the categories are "consumer protection," "hot new topics," "factors affecting weight loss," "dieting for events," and several medical topics related to dieting.

## Calorie Restriction (CR) Society
## <http://www.calorierestriction.org>

The CR Society is a nonprofit organization advancing the practice of caloric restriction as a means to enhance health and longevity. The Web site hosts e-mail discussion groups, encourages scientific research on CR, and educates the public about the diet. CR practitioners restrict caloric intake while consuming an adequate supply of vitamins, minerals, and other nutrients.* CR followers avoid simple sugars and flour, while consuming nutrient-dense foods, primarily vegetables. Fats and proteins are chosen carefully and in moderate amounts.

---

*Also known as caloric restriction with adequate nutrition (CRAN). Roy L. Walford, MD, did research on the concept while at the University of California at Los Angeles. In the mid-1980s and mid-1990s he published *The 120 Year Diet and the Anti-Aging Plan*. In 2005, Walford's daughter, Lisa Walford, and Brian M. Delaney, president of the Caloric Restriction Society, published *The Longevity Diet*. The motto is "fewer calories, more life."

## Carbohydrate Addict's Official Web site
<http://www.carbohydrateaddicts.com>

This is the Web site of Drs. Richard and Rachael Heller, whose books, including *The Carbohydrate Addict's Cookbook,* have long been popular. Carbohydrate addiction is defined as the overwhelming craving for carbohydrate-rich foods, caused by the body's excessive release of insulin (hyperinsulinemia). The carbohydrate addict gains weight easily and has swings in blood-sugar levels. The most useful part of this site is the FAQ section, but the Web site appears to be designed, primarily, to sell the Hellers' books.

*See also* **Atkins Nutritionals; South Beach Diet Online; The Zone Diet** (in this chapter).

## Cyberdiet @ DietWatch
<http://cyberdiet.com>

DietWatch, which acquired Cyberdiet in 2000, is the largest Internet diet site. Paid subscribers to CyberDiet receive a customized diet plan, meal plans, access to the online chat group, progress reports, shopping lists, fitness plans, and information on diet and nutrition. The site has won several awards from prestigious sources, including *Forbes, PC World,* and About.com. Timi Gustafson, a registered dietitian, and Cynthia Fink developed CyberDiet in 1995.

## Diet Center
<http://www.dietcenter.com>

The Diet Center is a franchise emphasizing close, one-on-one supervision, dietary supplements, and meal-replacement bars and drinks. The company has been in existence for over thirty years, so the program must succeed with some dieters.

## The Diet Channel
<http://www.dietchannel.com>

This site won *Forbes Magazine*'s Best of the Web award in the area of diet and nutrition. This is the place to go for brief descriptions of over sixty diets. It covers the spectrum from traditional (Weight Watchers) to faddish (Chocolate Diet) to bizarre (Russian Air Force Diet). An interesting feature is the section offering links to articles on diet and health issues such as phytochemicals, diet and cancer, antioxidants, and fad diets.

## Dr.Phil.com
## <http://www.drphil.com>

Weight loss is just one facet of Dr. Phil McGraw's Web site. Select "Weight" from the listing of topics. It's a highly commercial site, showing ads for his books and television show throughout. Weight-control advice stresses behavioral modification and understanding of the psychological aspects of overeating. Unique features are the dieting readiness profile, the "personal environmental audit," and "behavioral audit."

## Dr. Sears Zone Diet
*See* **The Zone Diet** (in this chapter).

## eDiets
## <http://www.ediets.com>

This is a commercial portal to two dozen well-known diet plans. ("Your diet, your way.") Sign in, select a diet from the list, indicate a goal weight and a target date, fill in some basic facts (name, address, e-mail address, height, weight, gender, and age), bypass or choose to receive various e-newsletters and product offers, and receive a free diet analysis. Visitors immediately receive (online) a graph showing current weight and the target date and weight, as well as a daily caloric guideline. Another graph shows BMIs and explains, for example, the health consequences if the BMI level is in the overweight or obese category. At this point, a registered dietitian presents the option of purchasing an online diet plan for $2.95/week, cancelable at any time by calling an 800 number. Note that there is a cancellation fee if cancelled during the first three months. Optional add-ons are available, such as professional support and access to Bob Greene's fitness training plan, which can significantly increase the weekly fee. Be careful to read the fine print explaining how the fees are charged to a credit card. The basic membership fee includes tracking progress, shopping lists, dining out and fast-food advice, recipes, and menus. If a dieter needs a lot of personal attention, but cannot meet in person with a diet counselor, eDiets may be the way to go.

## Healthy Weight Network
## <http://www.healthyweight.net>

Be sure to read the excellent tips in the section "How to Identify Weight Loss Fraud." The guidelines are extensive. There is also a section on how to report fraud. The site lists several questionable weight-loss pills, gadgets, and theories (although some might take exception to their dismissal of hypnosis).

## Herbalife
## <http://www.herbalife.com>

Herbalife is a twenty-five-year-old company that sells weight-loss and nutrition/wellness supplements and meal-replacement products through a network of independent distributors/ShapeWorks coaches. In order to purchase any of the products described on the Web site, fill out a form (listing a telephone number is mandatory) to be referred to a local distributor. The prices are not posted on the site. Visitors are able to view names of members of Herbalife's scientific and medical advisory boards.

## Jenny Craig
## <http://www.JennyCraig.com>

Known for its commercials featuring celebrity endorsements, this twenty-year-old franchised diet program emphasizes personal counseling and prepackaged, portion-controlled meals ("Jenny's Cuisine"). The Web site offers free recipes, a weight tracker, and advice, but consumers must sign up to receive the information. Jenny Craig also participates in eDiet's premium (fee-based) online diet program. Jenny Craig diet centers are located ("In-Centre [sp] Programs") in the United States, Canada, Puerto Rico, Australia, and New Zealand. Jenny Direct delivers prepackaged meals to your home and offers weekly telephone counseling and an online support group.

## Juice Fasts
## <http://altmedicine.about.com>

On this site select "Juice Fasting" and "How to Do a Juice Fast" under "Articles and Resources." Juice fasting consists of drinking only juices for a set period of time. Proponents claim the regimen detoxifies the body. Fasting is controversial and should never be attempted without discussing it beforehand with your physician or nurse. Please read the site's section "Who Should Not Try Juice Fasting."

## LA Weight Loss Centers
## <http://www.laweightloss.com>

The LA Weight Loss Company began in 1989. Today there are over 700 franchises worldwide. The Web site gives few cues as to the details of the weight-loss program. LA Weight Loss sells optional prepackaged foods, there are no group meetings or calorie counting. Members do not need to make appointments to come in for one-on-one counseling. Users are encouraged to visit a center or call an 800 number to enroll. Special features on the site are reserved for members. Free features include recipes, a BMI calculator, and an article from the American Heart Association offering ten tips for adding exercise to your daily routine.

## MayoClinic.com. Popular Diets: The Good, the Fad, and the Iffy
## <http://www.mayoclinic.com>

Select "Food & Nutrition Center" to read an article offering sound tips on evaluating popular diets. Note the "red flags" indicating a fad diet.

## Medifast
## <http://www.medifast1.com>

The company now known as Medifast began in 1972 as the Nutrition Institute of Maryland. At one time, the Medifast protein-sparing fast was only available under close medical supervision in hospitals or doctors' offices. Now, most clients purchase the Medifast products directly from the company. The very low-calorie diet consists of soy-based shakes and other meal-replacement foods that are consumed along with one "lean and green" daily meat and vegetable meal. Separate programs are offered for men and women, and a diabetic weight-loss program is available. The products meet kosher and lactose-free requirements.

## National Heart, Lung, and Blood Institute Obesity Education Initiative
## <http://www.nhlbi.nih.gov/health/public/heart/obesity/lose_wt/patmats.htm>

This site offeres key recommendations for the public stressing medical benefits to weight loss, the correct way to diet, and the importance of physical exercise.

## The Obesity Society (NAASO)
## <http://www.naaso.org>

This is the leading U.S. organization for the scientific study of obesity. It maintains statistics on this serious health problem and promotes advocacy, research, and education on the topic. It keeps physicians, scientists, and the public aware of new developments in the treatment and prevention of obesity. Consumers should read the organization's fact sheets on obesity, childhood obesity, obesity and cancer, and obesity and diabetes.

## NutriSystem Nourish
## <http://www.nutrisystem.com>

NutriSystem has been around for over thirty years, but the program is now Web-based. Brick-and-mortar NutriSystem diet centers no longer exist. Members purchase meal-replacement products (packages for men, women, type 2 diabetics, and vegetarians) through the Web site and receive counseling online or via the telephone. Membership is free and includes access to online chat groups, a newsletter, online classes, and unlimited counseling. Meal-replacement products are supplemented with vitamins and fresh fruit and vegetables. The NutriSystem meals include "good carbs"—carbohydrates with a low glycemic index. These carbohydrates are metabolized slowly and do not cause spikes in insulin levels.

## Optifast Medical Weight Loss Program
## <http://www.optifast.com>

Optifast, a medically supervised program produced by Novartis Medical Nutrition, is geared toward the clinically obese patient. A special regimen is available for bariatric surgery (weight-loss surgery) patients, both before and after surgery. Optifast formula was first developed in 1974. The program is offered only through hospitals and physicians' offices. Consumers can get local referrals on the Web site. A convenient feature of the site is the "toolbox" which pulls together links to nutrition, wellness, goal setting, fitness, and journaling (keeping a diary to record feelings and motivations).

## Overeaters Anonymous (OA)
## <http://www.oa.org>

Overeaters Anonymous is a twelve-step recovery program for compulsive overeaters. Group meetings are offered all over the world. Members pay no dues, assist new members, and help out at meetings. The site has a meeting locator. Read the "Twelve Concepts of Service" that forms the structure of OA. According to Naomi Lippel, managing director of OA, "OA is not just about weight loss, obesity, or diets; it addresses physical, emotional, and spiritual well-being. It is not a religious organization and does not promote any particular diet. . . . To address weight loss, OA encourages its members to develop a food plan with a health care professional and a sponsor."[5]

## Physicians Weight Loss Centers
## <http://www.pwlc.com>

This program of medically supervised diets started in 1979. Today there are five weight-loss systems and "in-center" members and "online" members. The five diet systems include programs for people needing to lose ten pounds or less, a more aggressive diet for those with larger weight-loss goals, a diet consisting entirely of Medifast meal-replacement products, a diet that is built around meals dieters prepare themselves, and a diet supplemented with weight-loss medications such as Meridia and Xenical. Franchised centers are located in approximately twelve states. Clients have blood work and licensed physicians perform EKGs, and there is one-on-one support. Unlike many other diet programs, there is careful monitoring of changes in body measurements. Sometimes the most dramatic changes are reflected in inches, not pounds.

## Richard Simmons
## <http://www.richardsimmons.com>

Richard Simmons has had a successful television show, a gym in Beverly Hills, California, exercise tapes, and diet aids (Deal-A-Meal and Food Mover). The Web site invites readers to join the "Clubhouse" in order to participate in online chats and other support options. The online store sells tapes and other Simmons products. The diet is based on motivation, exercise, and the American Dietetic Association/American Diabetes Association food-exchange system.

## Safediets.org
## <http://www.safediets.org>

This is the weight-loss diet site of Physicians Committee for Responsible Medicine (http://www.pcrm.org), a nonprofit organization supported by physicians, nutritionists, and laypeople who advocate preventive medicine and research, ethical treatment of human research subjects, alternatives to animal experimentation, and a low-fat, vegetarian diet. This program has a very detailed three-week weight-loss plan with meals built around four food groups: legumes, whole grains, fruits, and vegetables. When the diet fact sheet is printed out and duplicated, it makes a very convenient daily diary for checking off required servings of the food groups as they are consumed each day. There is also a section rating the top diet books (but at the time of this writing, it does not mention the popular *South Beach Diet* book).

**Shape Up America**
*See* **Shape Up America!** (in Chapter 5).

**Sisters Together: Move More, Eat Better**
**<http://win.niddk.nih.gov/sisters/index.htm>**

Sisters Together, a program of the National Institutes of Health Weight-Control Information Network, features several online publications on its Web site: *Active At Any Size, Nutrition and Your Health, Weight Loss for Life, Sisters Together Program Guide,* and *Improving Your Health: Tips for African American Men and Women,* and other materials from the program.

**Slim-Fast.com**
**<http://www.slim-fast.com>**

Slim-Fast offers four versions of the Optima Diet, combining meal-replacement shakes and bars with sensible meals and snacks. Three of the plans are geared to the dieter's starting weight, while the fourth option is a low-carb diet. The products are carried in supermarkets and drug stores. The Web site offers lots of dieting advice, online chats, meal planning guidelines, and recipes.

**Society for Clinical and Experimental Hypnosis**
**<http://www.sceh.us>**

SCEH is an international organization founded in 1949. Its members are health care professionals researching and practicing the medical use of hypnosis. SCEH publishes the *International Journal of Clinical and Experimental Hypnosis,* a respected scholarly journal available in many medical school libraries. The organization offers workshops and symposia for health care practitioners wishing to further their hypnotherapy skills. Consumers can contact the society for a list of professionals.

**South Beach Diet Online**
**<http://www.southbeachdiet.com>**

Miami cardiologist Arthur Agatston, MD, developed the South Beach diet for his heart patients. He has since published best-selling books on the diet. The South Beach Diet is built around vegetables, some fruit, low-fat protein, olive and canola oils, and carbohydrates with low glycemic-index ratings, such as whole-grain breads and pasta. There are three phases to the diet, with carbohydrates and fruit added gradually during the second phase. Members, who are billed quarterly, have access to Q&A, chats, a weight tracker, meal plans, an online journal, message boards, Beach Buddies (dieters matched to other members for support), advice from online dietitians, food guides, and over 900 recipes.

*See also* **Atkins Nutritionals; Carbohydrate Addict's Official Website; The Zone Diet** (in this chapter).

## Sugar Busters! Diet
**<http://www.sugarbusters.com>**

The mantra of this program is "cut sugar to trim fat." Three physicians, Samuel S. Andrews, Morrison C. Bethea, and Luis A. Balart, and a successful follower of the diet, H. Leighton Steward, wrote the *Sugar Busters!* book. The concept of the diet is that consumption of refined sugar and bad carbohydrates (processed grain products) causes the body to produce more insulin, store fat, and gain weight. On the Sugar Busters! program, dieters are encouraged to eat fruit, high-fiber vegetables, and low-fat meats and fish. Exercise is an important component of the program. The Web site sells frozen meals, snacks, beverages, and sugar substitutes.

## Trevose Behavior Modification Program (TBMP)
**<http://www.tbmp.org>**

Shape up or ship out! A tough-love attitude may be what a dieter needs to finally get serious about weight loss. Insurance executive David S. Zelitch, under the direction of world-famous obesity authority Dr. Albert J.Stunkard, MD, of the University of Pennsylvania School of Medicine, started TBMP in Philadelphia in 1970. In order to apply to join the program, check the Web site for satellite meeting groups that have openings. Once there is an opening, send a letter stating what has been done in the past to lose weight and why you believe you are ready to lose weight now. Include a self-addressed, stamped envelope. Most members wait three to six months before they are admitted into the program. There are no fees, but participants are encouraged to make contributions to cover the cost of supplies. A newsletter is available. Participants are required to lose a predetermined amount of weight each month. They must attend every weekly meeting. The emphasis is on behavior modification, and the first six months on the program are devoted to studying participants' eating habits. Read the "Rules" section of the Web site to learn all the weight-loss and attendance requirements. The program offers satellite meetings only in the Delaware Valley (southeastern Pennsylvania and southern New Jersey). Successful participants are encouraged to give back to TBMP by becoming trained meeting leaders. Anyone who flunks out of the program is not permitted to reapply.

## The V8 Diet Plan
## <http://www.v8juice.com>

The V8 Diet Plan is a balanced diet built around V8 vegetable juice and V8 Splash combined with sensible meals. There is a diet plan and a chart with suggested substitutions (drink a Splash instead of a frappuccino and save 180 calories).

## WebMD Weight-Loss Clinic
## <https://diet.webmd.com>

On this site, a nutritionist customizes an eating plan based on personal weight-loss requirements and current eating habits and food preferences. There is a weekly fee, charged quarterly to a credit card. Members have access to an online community, and an interactive journal for tracking food consumption and weight-loss progress.

## The Weigh Down Workshop
## <http://www.wdworkshop.com>

Gwen Shamblin, a registered dietitian and university nutrition instructor with experience as a weight-loss consultant, founded this Christianity-based diet program in the 1980s. Members purchase syllabi for courses that are offered online or on DVDs. Members who recruit five or more participants receive a free copy of the syllabus.

## WeightWatchers.com
## <http://www.weightwatchers.com>

The Weight Watchers Web site has three components: (1) free information about the diet programs and traditional meetings, a message board, a community recipe exchange, and a few free Weight Watchers recipes; (2) Weight Watchers Online, a subscription-only resource; and (3) Weight Watchers eTools, a subscription-only Internet companion to traditional meetings.

Traditional Weight Watchers meetings are held all over the world, and the Web site makes it easy to locate local meetings. The diet program regularly adds innovative enhancements (new ways to track flexible points and a "no counting" plan option), and the techniques used by meeting instructors have changed to encourage more member participation through meal planning exercises and questions presented for group discussion.

## Weight-Control Information Network (WIN)
## <http://win.niddk.nih.gov/index.htm>

WIN, part of the National Institutes of Health, produces, collects, and disseminates information targeted at consumers and health professionals on different aspects of obesity, weight control, and nutrition. In addition to brochures and fact sheets (available online), WIN produces media programs, and has established Clinical Nutrition Research Centers (CNRCs) and Clinical Nutrition Research Units (CNRUs) to advance obesity and nutrition research. Be sure to read *Choosing a Safe and Effective Weight-loss Program* and *Diet Myths* (see Figure 6.2).

## The Zone Diet
## <http://www.zonediet.com>

The Zone diet, developed by former Massachusetts Institute of Technology biotechnology researcher Dr. Barry Sears, consists of eating moderate amounts of carbohydrates, protein, and fats at every meal and snack. The Zone's proportions for these nutrients are 40-30-30, with

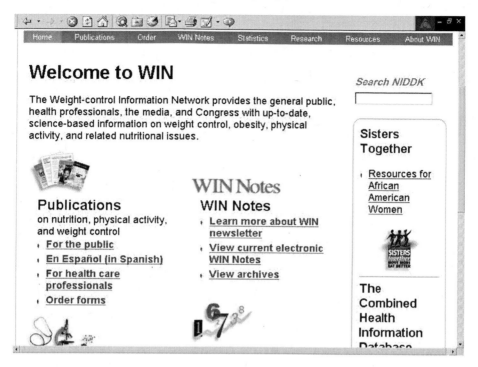

FIGURE 6.2. Weight-Control Information Network
<http://www.niddk.nih.gov/index.htm>

the goal of keeping insulin levels even. Dr. Sears' book *The Zone* was published in 1995. Prepackaged meals and supplements are sold at the Web site, but they are not a required part of the diet. Members receive access to support, a diet journal, an online community, tips, and customized shopping lists and recipes.

*See also* **Atkins Nutritionals; Carbohydrate Addict's Official Website; South Beach Diet Online** (in this chapter).

## NOTES

1. Lissau, I., Overpeck, M.D., Ruan, W.J., Pernille, D., Holstein, B.E., Hediger, M.L., and the Health Behaviour in School-Aged Children Obesity Working Group (2004). Body Mass Index and Overweight in Adolescents in European Countries, Israel, and the United States. *Archives of Pediatrics and Adolescent Medicine* 158: 27-33.

2. Sturm, R. (2002). The effects of obesity, smoking, and drinking on medical problems and costs. *Health Affairs* 21: 245-253.

3. Finkelstein, E.A., Fiebelkorn, I.C., and Wang, G. (2004). State-level estimates of annual medical expenditures attributable to obesity. *Obesity Research* 12: 18-24.

4. Alpert, J.S. and Powers, P. J. (2005). Editorial: Obesity: A complex public health challenge. *American Journal of Medicine* 118: 935.

5. Text added at request of OA's managing director. Personal e-mail correspondence between the author and Naomi Lippel, December 27, 2005.

# Chapter 7

# Weight Loss (Bariatric) Surgery Web Sites

Bariatric (*baros* is the Greek word meaning "weight" and *iatrikos* meaning "art of healing") surgery involves reducing the capacity of the stomach by means of one of several procedures. In 2003, an estimated 100,000 people had bariatric surgery, and the number is expected to exceed 145,000 in 2004. Of American adults, 6 to 10 percent are considered to be morbidly obese, having a BMI greater than 40. Generally, bariatric surgery is restricted to people in this category. People with major, life-threatening diseases (comorbidity factors), such as poorly controlled hypertension (high blood pressure) or diabetes, severe heart/lung problems, or obstructive sleep apnea may also be candidates for the surgery, even though they have lower BMIs (35-40).

A psychological evaluation is usually part of the presurgical process. The prospective patient must understand the risks of bariatric surgery. In the United States, one of every 200 people who undergo this surgery dies within thirty days of the procedure. Serious complications following surgery include lung blood clots (pulmonary emboli), wound infections, bleeding, and gastrointestinal leakage. The benefits of this major abdominal surgery must exceed the risks. Patients' responsibilities do not end with the surgery: they must commit to a lifetime of careful monitoring of their food intake, general health, and vitamin and mineral status (anemia and calcium, iron, or vitamin $B_{12}$ deficiencies are common). Patients are cautioned to avoid foods high in fat or refined sugar, lest they develop "dumping syndrome" (a group of unpleasant symptoms including dizziness, nausea, diarrhea, and rapid heart beat).

*Internet Guide to Medical Diets and Nutrition*
© 2006 by The Haworth Press, Inc. All rights reserved.
doi:10.1300/5852_07

After bariatric surgery, the stomach is much smaller and may hold only 2 to 6 ounces of food at one time (a normal stomach holds 40 to 50). During meals patients are cautioned to avoid drinking fluids and to chew food until almost liquid. Dry foods, such as nuts, bread, and popcorn may stick and cause vomiting. High-fiber foods can cause cramping. There may be cosmetic issues to resolve after losing a great deal of weight: sagging skin on the face and body. Some patients will require plastic surgery to deal with hanging folds of skin on the arms and abdomen. Late-term complications, such as intestinal blockage due to rearrangement of the intestines during surgery, have occurred. Whenever someone makes a major change, issues may arise with family members and friends. The patient may have problems adjusting to his or her new appearance or may continue to view himself or herself as obese (body-image disturbances). The actual weight loss or change in appearance may not be what was expected. Mattison and Jensen report "patients must realize that the definition of success for these procedures is the loss of 50 percent of their excess weight . . . it seldom results in the patient achieving his or her ideal weight."[1] The weight will come back if the patient does not exercise and eats small amounts of fattening food all day long ("outeating").

In the United States, the most popular types of bariatric surgical procedures include laparoscopic gastric banding (Lap-Band adjustable gastric band), including vertical banded gastroplexy (VBG), and gastric bypass (Roux-en-Y gastric bypass), representing the categories "restrictive" and "malabsorptive," respectively. Currently being evaluated is a safer, two-stage procedure in which patients first undergo a laparoscopic sleeve gastrectomy, in which a large portion of the stomach is removed. Months later, they undergo the Roux-en-Y procedure that involves creation of a small stomach pouch and a bypass of a portion of the intestines.[2,3,4] Procedures cost an average of $25,000. The following Web sites offer detailed information on the specific procedures.

## American Society for Bariatric Surgery
## <http://www.asbs.org>

This professional organization of physicians specializing in weight-loss surgery was established in 1983. Its mission is to educate and train physicians, educate the public and allied health professionals, set guidelines for the ethical selection and treatment of patients, exchange knowledge and information among its members, conduct research, and promote quality assurance and evaluation of surgical outcomes. The Web site features a BMI calculator, lists of continuing medical education courses for physicians, "Rationale for the Surgical Treatment of Morbid Obesity," and a lengthy history of the evolution of bariatric surgery.

## Gastrointestinal Surgery for Severe Obesity
## \<http://win.niddk.nih.gov/publications/gastric.htm\>

Part of the Weight-Control Information Network (WIN) of the National Institutes of Health National Institute of Diabetes and Digestive and Kidney Diseases. WIN was established in 1994 to provide science-based information on obesity, weight control, and nutrition to consumers and health professionals. This is one of the most extensive Web resources on bariatric surgery. On the site are illustrations, very clear and thorough descriptions of the advantages and drawbacks of all types of gastric bypass surgery, and a list of questions a patient should ask himself or herself before deciding to undergo this surgery.

## MedlinePlus.gov
## \<http://medlineplus.gov\>

Select "Surgical Videos" to view videotapes of bariatric surgical procedures. Not for the squeamish.

## ObesityHelp.com
## \<http://www.obesityhelp.com/morbidobesity\>

In 1998, Eric Klein, a former emergency medical technician, founded the Association for Morbid Obesity Research and the Web page ObesityHelp to provide peer support for the morbidly obese, including patients undergoing or contemplating bariatric surgery. This site is somewhat commercial, but there are so many informative sections, that it should be a top selection for patients, no matter at what stage they are dealing with obesity issues. On the site are a clothing exchange, "Ask a Surgeon," "Find a Bariatric Surgeon," "Find a Plastic Surgeon," and "Find a Hospital." The "Bariatric Hospital Directory" is very useful: click on a state and view information about local facilities that perform this type of surgery. The directory lists number of procedures performed, date of the most recent procedure, accreditation of the hospital, and detailed evaluations by former patients. Patients can post personalized profiles and locate peers. Patients may elect to list their name, photograph, type of surgery, and date of scheduled surgery ("Recent and Upcoming Surgeries"). Peers are encouraged to provide support, especially the week before and after the surgery. Reading the diary-like entries, especially the material about the postoperative period, should be of immense help to people contemplating weight-loss surgery. Several types of chat rooms ("forums") are available, including one on nutrition, with information contributed by dietitians. The detailed descriptions of several types of surgical procedures is excellent, with illustrations and advantages/disadvantages displayed as a chart. Klein requests contributions and hospital/physician sponsorship to help defray costs; you can subscribe to a print magazine *(ObesityHelp)*. Over 150,000 people have been members over the years.

ObesityHelp is the most valuable Web site in this category because of the accounts of patient experiences (see Figure 7.1).

## Obesity-Surgery.Net
**<http://www.obesity-surgery.net/obesity_surgery_faq.htm>**

This is the Web site for a bariatric surgeon's clinic, but the FAQ section is clearly and honestly written. Note the realistic information on weight loss and regain of lost weight after bariatric surgery.

## Obesity Surgery Support Group (OSSG)
**<http://groups.yahoo.com/group/ossg>**

OSSG, also called Planet OSSG, is a moderated discussion group for people who have had bariatric surgery or are contemplating having this surgery. To subscribe, read the instructions on this site.

FIGURE 7.1. ObesityHelp
<http://www.obesityhelp.com/morbidobesity>
Reprinted with permission of ObesityHelp.com.

## Thinner Times Gastric Bypass
## <http://thinnertimes.com>

Thinner Times is the Web site of a private surgical clinic, but the information about gastric surgery and the links to support groups are useful. Be sure to read the section on complications (select "Gastric Bypass" on the home page).

## NOTES

1. Mattison, R. and Jensen, M.D. (2004). Bariatric surgery: For the right patient, procedure can be effective. *Postgraduate Medicine* 115: 49-58.
2. Gallagher, S. (2004). Taking the weight off with bariatric surgery. *Nursing* 34: 58-63.

3. Steinbrook, R. (2004). Surgery for severe obesity. *New England Journal of Medicine* 350: 1075-1079.

4. Windham, C. (2003). "Gastric-bypass study shows risks of surgery," *Wall Street Journal,* December 2, p. D6, eastern edition.

Chapter 8

# Diet, Nutrition, and Recipe Web Sites for Medical Problems and Conditions

Feed a cold, starve a fever.

Anonymous saying

In addition to the following Web sites, note the downloadable fact sheets on the role of nutrition in preventing and managing medical conditions at Physicians Committee for Responsible Medicine (http://www.pcrm.org). In researching medical diets, if no authoritative sites were found, diets for a particular disease were not included in this chapter. If a particular type of diet was never recommended by an authoritative source, that type of diet was omitted from this chapter. *Note:* All Web sites are included for their information value only. We do not offer medical advice, and suggest you consult with your physician before trying any of these diets.

## *AIDS/HIV*

### San Francisco AIDS Foundation: AIDS 101: Guide to HIV Basics <http://www.sfaf.org/aids101/treatment.html>

Select "Nutrition" for detailed information on the high-protein, well-balanced diet required to sustain people with AIDS or HIV. The San

*Internet Guide to Medical Diets and Nutrition*
© 2006 by The Haworth Press, Inc. All rights reserved.
doi:10.1300/5852_08

Francisco AIDS Foundation, established in 1982, is one of the oldest community-based AIDS service organizations.

## ALZHEIMER'S DISEASE

### Alzheimer's Prevention Foundation International
### <http://www.alzheimersprevention.org>

Select "Prevention" and then "Pillar 1—Diet and Vitamins." Dr. Dharma Singh Khalsa's Alzheimer's Prevention Diet is outlined here. Dr. Khalsa, a physician, is president and medical director of the foundation.

## ANEMIA

### WomensHealth.gov/Anemia
### <http://www.4woman.gov/faq/anemia.htm>

The site is careful to distinguish among the many causes of anemia. The nutrition recommendations are geared toward iron-deficiency anemia.

## ARTHRITIS

### Arthritis Foundation
### <http://www.arthritis.org>

The Arthritis Foundation is a national, nonprofit group funding research, providing advocacy, and promoting public education about the many different types of arthritis. The disease affects one out of six people in the United States. Select "Resources," then "Diet and Nutrition." *Diet and Your Arthritis* can be purchased from the Web site. A free summary of the brochure is also on the site.

## *ATHEROSCLEROSIS OR HIGH CHOLESTEROL*

*See also* HEART DISEASE (in this chapter).

### HeartPoint Health Information You Can Trust
### <http://www.heartpoint.com>

This is the Web site of cardiologist, Darrell J. Youngman, DO, who is board certified in both internal medicine and cardiology. Read the sections on the low-fat diet and "Foods You Will Love."

### The Ornish Heart-Healthy Lifestyle Program (on WebMD)
### <http://www.ornish.com>

California's Dean Ornish, MD, has developed two diets, one for *preventing* heart disease and high cholesterol, and the other for *reversing* these diseases. Both are part of the Ornish Lifestyle Program, now residing on the WebMD Web site. Arteriosclerosis goes hand in hand with heart disease and long-standing high cholesterol. See also Dr. Dean Ornish Program for Reversing Heart Disease at Windber Medical Center (http://www.windbercare.com/ornish.html).

### Therapeutic Lifestyles Changes Diet/TLC Diet
### <http://nhlbisupport.com/cgi-bin/chd1/step2intro.cgi>

This site contains National Heart, Lung, and Blood Institute suggestions for implementing a diet designed to lower "bad" cholesterol—LDL (low-density lipoprotein) level.

### WomensHealth.gov. Heart Healthy Eating.
### <http://womenshealth.gov/faq/healtheat.htm>

WomensHealth.gov is the federal government's resource for health information targeted toward women's medical concerns and unique requirements. Heart disease is a major killer of women. This Web site discusses and provides links to the Therapeutic Lifestyles Changes (TLC) Diet aimed at lowering cholesterol, the Heart Healthy Diet, which also can lower cholesterol, and to the DASH Diet, which lowers

FIGURE 8.1. WomensHealth.gov Heart Healthy Eating
<http://womenshealth.gov/faq/healtheat.htm>

high blood pressure. Site contains a useful chart comparing the sodium/fat/caloric limits for the TLC and Heart Healthy Diets (see Figure 8.1).

## ATTENTION-DEFICIT/HYPERACTIVITY DISORDER (ADHD)

### Ask the Dietitian/Hyperactivity and ADHD
### <http://www.dietitian.com/hyperactive.html>

Registered dietitian Joanne Larsen debunks some of the research promoting the Feingold elimination diet for children with Attention Deficit/Hyperactivity Disorder.

## Children and Adults with Attention-Deficit/Hyperactivity Disorder (CHADD)
### <http://www.chadd.org>

CHADD supports individuals, encourages research, advocates for legislation, and educates the public and physicians by publicizing evidence-based scientific research on ADHD. Note the information on lack of research evidence supporting elimination of food additives and preservatives (Feingold diet) or diets that are free of sugar and candy. CHADD says there is no definitive research to suggest supplementing the diet with megadose vitamins, herbal preparations, fatty acids, single-dose vitamins, or amino acids.

## *AUTISM*

## Autism Society of America
### <http://www.autism-society.org>

The Autism Society of America is an organization composed of parents, teachers, therapists, and physicians. It was started in 1965 by Bernard Rimland, PhD, and now has over 50,000 members, with 200 chapters nationwide. Through its advocacy efforts, legislation has been passed protecting the rights of people with autism and mandating their education. Select "Treatment" and read "Biomedical and Dietary Approaches."

## *CANCER (INCLUDING CHEMOTHERAPY, CANCER PREVENTION, POSTRADIATION RECOVERY)*

*See also* LEUKEMIA (in this chapter).

## American Cancer Society
### <http://www.cancer.org>

On this site, select "Patients, Family, and Friends," then "Coping With Treatment" to see the link "Nutrition for Cancer Patients." In this section are subtopics such as "When Treatment Causes Eating problems," "Nutrition for Children with Cancer," and recipes.

### American Institute of Cancer Research (AICR)
<http://www.aicr.org>

AICR, founded in 1982, supports research into the role of diet in the treatment and prevention of cancer. Read the advice in "AICR's Diet and Health Guidelines for Cancer Prevention."

### Cancer Project of the Physicians Committee for Responsible Medicine
<http://www.cancerproject.org>

Physicians and nutritionists collaborate to educate the public and conduct clinical research on the role of a healthy (vegetarian) diet in "cancer prevention and survival." Find the location of free cooking and nutrition classes, and read about cancer protective foods and nutrients.

### Memorial Sloan-Kettering Cancer Center
<http://www.mskcc.org>

Select "Prevention and Screening," then "Nutrition Counseling" for advice on how to follow a healthy diet. Laura Pensiero, a registered dietitian and cookbook author answers questions and provides nutritional guidelines.

### National Cancer Institute. Eating Hints for Cancer Patients: Before, During, and After Treatment
<http://www.cancer.gov/cancerinfo/eatinghints>

This is an excellent site because it is thorough and written in easy-to-understand language. The tone is reassuring, emphasizing the need to adopt a positive attitude. There is an emphasis on the importance of consulting with a registered dietitian, and the Web site provides the telephone number of the free hotline of the American Dietetics Association.

Throughout the text are links to a glossary of words a patient might not understand. Recipes and practical advice are offered on stocking a refrigerator ahead of time with favorite foods. Nutrition for cancer patients is quite different from nutrition for healthy patients. Cancer patients may need to eat high-calorie foods to boost stamina. The resource lists possible side effects of treatment that may effect appetite or the ability to eat; there are suggestions for coping with each difficulty.

## CELIAC/COELIAC DISEASE, NONTROPICAL SPRUE, GLUTEN INTOLERANCE, GLUTEN-SENSITIVE ENTEROPATHY

Celiac disease is a lifelong genetic autoimmune disorder often characterized by diarrhea and malabsorption brought on by eating foods containing the protein gluten. The disease is triggered by intolerance to gluten. Increased antibodies to gluten protein (primarily *tissue transglutaminase*) form in the small intestines, injuring the brush border, which prevents normal absorption of minerals, vitamins, and other nutrients. People with celiac disease may also suffer from other autoimmune diseases such as type I diabetes, rheumatoid arthritis, lupus, or Sjogren's disease. Some patients have early-onset osteoporosis or unexplained fatigue or anemia. People with celiac disease must avoid barley, rye, wheat, and possibly oats. The disease affects an estimated 1 in every 120 to 300 people in Europe and North America. It is estimated that 2 million people in the United States suffer from the condition. There are no medications or surgical procedures to treat this disease, so patients must strictly follow a gluten-free diet for life.[1] Failure to avoid gluten in the diet can lead to malnutrition and other serious conditions. Patients with celiac disease are more likely to develop diabetes, epilepsy, some types of cancer, dermatitis, and other diseases.

### Alt.Support.Celiac

This is an unmoderated newsgroup.

## American Celiac Society
## <http://www.americanceliacsociety.org>

The American Celiac Society, based in New Orleans, has resumed normal operations since Hurricane Katrina. Annette and James Bentley run the 501(c)(3) nonprofit organization, which, through its Dietary Support Coalition, focuses on education of patients and the public about celiac disease, offers support, sponsors conferences, supports research, and identifies ingredients in foods and food supplements.

## Canadian Celiac Association
## <http://www.celiac.ca>

The site may be viewed in English or French. The food alerts section warns of contaminated food products. The gluten-free diet information is very thorough, with much of the information applicable to consumers living in the United States.

## Celiac Canada
## <http://www.penny.ca/Celiac-canada.htm>

Celiac Canada is a mailing list and a Web site. The site might be more appealing to teens because there is a section on famous people with celiac disease (mostly rock stars and actors) and a fast- food guide.

## Celiac.Com: Celiac Disease and Gluten-Free Diet Support Center
## <http://www.celiac.com>

On this site you can purchase the gluten-free dining guide to get information on safe dining at over seventy restaurant chains. Print out free recipes from an extensive list. The FAQs section gives reliable, thorough information from a celiac-L mailing list.

## Celiac Disease Foundation
## \<http://www.celiac.org\>

This site includes product alerts and a list of potentially harmful in-gredients. The foundation sponsors many social events and has support groups ("Connections").

## Celiac-L (Celiac/Coeliac Wheat/Gluten-Free List)
## \<http://lists.wikicities.com/mailman/listinfo/celiac-1\>

Subscribe to this moderated mailing list by going to this Web site and filling in the application form.

## Celiac Sprue Association
## \<http://www.csaceliacs.org\>

The Celiac Sprue Association is a nonprofit, member-based support organization. The Web site is very well designed and easy to navigate. It bears the HONcode seal of approval. The association is particularly strong on advocacy, and actively solicited member input to assist the Food and Drug Association's project to standardize gluten-free food product labeling. Select "Recipes" and "Gluten-free Diet" for help in planning nutritious, safe meals. A separate section for children and teens includes recipes children can prepare.

## Club Celiac
## \<http://www.clubceliac.com/clubceliac.htm\>

Lucy Shriver, a patient with celiac disease, created and maintains this Web site. She also maintains The Gluten Free Kitchen/Cooking with Lucy (http://www.glutenfreekitchen.org). Children can exchange reci-pes here, purchase a gluten-free cookbook, read the message board, or join the chat line.

## Gluten Intolerance Group of North America
## \<http://www.gluten.net\>

The Gluten Intolerance Group is a nonprofit organization that advo-cates for legislative reform, sponsors children's camps, and publishes educational materials. You can download *Quick Start Diet Guide for Celiac Disease*. The site also contains a brief guide to the gluten-free diet.

## Gluten-Free Mall
## <http://www.glutenfreemall.com>

This is a commercial site that sells gluten-free products. Search for products by ingredient or category. Includes frozen foods.

## National Digestive Diseases Information Clearinghouse:
##   Celiac Disease
## <http://digestive.niddk.nih.gov/ddiseases/pubs/celiac/index .htm>

Select "What is the Treatment" to read an overview of the diet and see a chart with examples of meals, including recommended amounts.

## National Foundation for Celiac Awareness (NFCA)
## <http://www.celiacawareness.org>

This nonprofit organization encourages research to find a pharmaceutical cure for celiac disease. It funds programs to raise awareness about the disease and support patients, and promotes screening programs. NFCA's site includes the latest news about celiac disease, personal stories of patients, and links to other resources on the Web. The Web site has a very thorough explanation of the disease process, the symptoms, and how the disease is treated.

## *CHEWING/SWALLOWING DISORDERS*

## Dinner Through a Straw [recipes]
## <http://www.dinnerthroughastraw.net>

This is a commercial Web site selling devices to enable people to eat liquids and liquefied foods. The products are marketed to patients with TMJ (temporomandibular joint dysfunction), scleroderma, cancer, AIDS, and following jaw surgery. The company also sells a recipe book, *Dinner Through a Straw,* written by an audiologist.

## ZNS Products
## <http://www.zip-n-squeeze.com>

Zip-N-Squeeze bags were developed by registered nurse Susan Beaudette. A free smoothie recipe is offered on the Web site.

## *COLITIS, CROHN'S DISEASE, INFLAMMATORY BOWEL DISEASE (IBD)*

## Crohn's and Colitis Foundation of America
## <http://ccfa.org>

This nonprofit organization educates the public, advocates for legislative change, funds research, and supports the sufferers of ulcerative colitis and Crohn's disease. Select "Living with IBD" and then "Diet & Nutrition"; within this section, there is a link to recipes.

## Mayo Clinic.com: Ulcerative Colitis
## <http://www.mayoclinic.com>

Select "Diseases & Conditions A to Z." Click "U" for ulcerative colitis. Select "Self-Care" and read the diet tips for patients who have the disease. Ulcerative colitis affects more than 500,000 people in the United States. Symptoms include watery diarrhea and severe abdominal pain. Foods do not cause the disease, but symptoms may be kept at bay by following certain dietary guidelines. The medical advisors at Mayo Clinic.com suggest avoiding known trigger foods, such as certain high-fiber fruits and vegetables, whole grains, high-lactose dairy products, colas, caffeine, alcoholic beverages, popcorn, broccoli, and beans. Patients should drink plenty of noncarbonated water. Several small meals a day may be easier to digest.

# COLOSTOMY

## Living with a Colostomy
## <http://www.ostomysupport.info/diet.html>

On this site, a patient gives tips on managing difficulties with foods postsurgery and thereafter. Most sites written by physicians suggest that no diet changes are necessary after the immediate postsurgery phase, but, in reality, patients need to learn which foods cause them to suffer excess gas or diarrhea.

# DEPRESSION

## Elizabeth Somer's Web site
## <http://www.elizabethsomer.com>

Elizabeth Somer is a registered dietitian who frequently appears on national television. Ask questions, read some of her articles, or purchase her book, *The Food and Mood Cookbook.*

## HealthyPlace
## <http://www.healthyplace.com>

HealthyPlace is an online community of people providing mental health information and support. Consumers participate in forums, chat groups, and bulletin boards. Medical journalists (not physicians or mental health professionals) write news articles on mental health topics. Select "Depression" under HealthyPlace Communities, click on "Alternative Remedies for Depression." Read the information under "Dietary Supplements and Nutrition for Depression."

# DIABETES MELLITUS

## American Diabetes Association
## <http://www.diabetes.org>

On this site you can purchase cookbooks and download free recipes and nutrition guides. This Web site is packed with authoritative information for living a healthier life with diabetes. Browse through two sections: "Weight Loss & Exercise" and "Nutrition & Recipes" (see Figure 8.2).

## *Diabetic Gourmet*
## <http://diabeticgourmet.com>

*Diabetic Gourmet* magazine is an online magazine, published since 1995. It is dedicated to "diabetic dining and healthy living." Select "Food and Dining" to read news items on food, cuisine, and nutrients for diabetics. Search the *Diabetic Gourmet* recipe archive. Be sure to check "Tools and Calculators" for some tools of specific interest to diabetics: weight, volume, and temperature calculators; a converter for blood-sugar readings; a tool to convert whole-blood readings to plasma

FIGURE 8.2. American Diabetes Association
<http://www.diabetes.org>
Reprinted with permission of the American Diabetes Association.

readings; walked-off-calories calculator; and a basal-metabolic rate (BMR) calculator.

## EverydayEating.com
## <http://www.everydayeating.com>

Nestlé USA produces this resource for people with diabetes. The site features recipes, extensive information about nutrition for diabetics, and Q&A. Nestlé also manufactures Lean Cuisine frozen foods, Carnation Instant Breakfast, and several sugar-free and low-carb Nestlé products. Read "Nutrition 101." Did you know that the glycemic index (GI) of fruits and vegetables can go up with ripeness or method of preparation? The GI can go down if the vegetables are served with high-fiber or high-fat foods, as in a casserole.

## Joslin Diabetes Center
## <http://www.joslin.org>

The world-renowned Joslin Clinic began in 1898 with a focus on the treatment of diabetes. Now it is affiliated with Harvard Medical School. There are Joslin Diabetes Center satellites and affiliates worldwide. Patients can purchase cookbooks from the online Joslin Store. Free nutrition information is available, including meal-planning advice and recipes in the section "Joslin's Library: Managing your Diabetes" (choose "Learn About Diabetes" on the home page).

## National Diabetes Information Clearinghouse
## <http://www.diabetes.niddk.nih.gov>

On this site, select "Treatments for Diabetes" to read meal-planning tips under "What I Need To Know About Eating and Diabetes." You can download, in English or Spanish, "Recipe and Meal Planner Guide" from the National Diabetes Education Program.

## DIGESTIVE DISEASES

*See also* REFLUX ESOPHAGITIS/GERD/HEARTBURN and specific disorders in MedlinePlus.gov (http://medlineplus.gov).

**National Digestive Diseases Information Clearinghouse**
**<http://digestive.niddk.nih.gov/>**

You can read material under treatment of specific diseases. The clearinghouse gives dietary recommendations.

## DIVERTICULITIS/DIVERTICULOSIS

*See also* **MayoClinic.com.**

**Go Ask Alice! Diverticular Disease (Diverticulosis) and Diet**
**<http://www.goaskalice.columbia.edu/1309.html>**

Go Ask Alice! is an Internet health question and answer service produced by Columbia University's Health Promotion Program. Over 3,000 Q&As are currently archived on the site. The information provided on diverticular disease is thorough, yet appropriate for consumers. The site includes links to related Q&As on the Web site.

## FIBROCYSTIC BREAST DISEASE

**MedlinePlus Medical Encyclopedia: Lumps in the Breasts**
**<http://medlineplus.gov>**

On this site select "Medical Encyclopedia," then use the A to Z directory to select "Lumps in the Breasts." MedlinePlus is a consumer health Web site produced by the National Library of Medicine. All material is from authoritative medical resources.

## *FOOD ALLERGIES*

### Food Allergy and Anaphylaxis Network (FAAN)
<http://www.foodallergy.org>

FAAN is a worldwide organization of physicians, patients, dietitians, and nurses. It was started in 1991 to raise awareness about these conditions, advocate for legislative change, and promote research. This site provides very extensive information for teens and adults living with food allergies. There are allergen alerts and recipes.

### Oregon Health and Science University (OHSU Health)—
Allergens: Food
<http://www.ohsuhealth.com/htaz/allergy/allabout_allergy/
allergens/foods/index.cfm>

OHSU is a medical school. The section on food allergies is written for consumers and contains excellent information on the following: diet for lactose intolerance; and diets for egg, milk, peanut, shellfish, soy, tree nut, and wheat allergy.

## *GLUTEN-FREE DIET*

*See* CELIAC DISEASE (in this chapter).

## *GOUT/LOW-PURINE DIET*

### MayoClinic.com: Gout Diet: Reducing Purines
<http://www.mayoclinic.com/invoke.cfm?id=HQ00765>

Physicians write the material on MayoClinic.com. The section on the gout diet explains new medications that may obviate the need for dietary restrictions.

The Web site warns that very restrictive weight-loss diets, such as some of the more popular low-carbohydrate diets, may cause or aggravate gout. Extreme dieting, diets high in protein or fat, organ meats, and some fish contribute to high levels of purine in the body, which break down to produce high levels of uric acid. This can cause gout.

## HEADACHES/MIGRAINE ELIMINATION DIET

*See also* MAOI DIET (in this chapter).

### Physicians Committee for Responsible Medicine. A Natural Approach to Migraines
### <http://www.pcrm.org/health/prevmed/migraine.html>

This site offers excellent advice from PCRM. Note the list of "pain-safe" foods that rarely trigger headaches, as well as the list of foods to eliminate ("common triggers"). The trigger foods contain high amounts of tyramine.

## HEART DISEASE/CONGESTIVE HEART FAILURE (INCLUDING TREATMENT AND PREVENTION)

*See also* ATHEROSCLEROSIS OR HIGH CHOLESTEROL and LOW-SODIUM DIET (in this chapter) and Chapter 10, "Weight Loss Spas and Residential Diet Programs."

### American Heart Association
### <http://www.americanheart.org>

On this site, read "Diet & Nutrition" under "Healthy Lifestyle." Select "American Heart Association's No-Fad Diet" to read excerpts from this book. The free brochure, *An Eating Plan for Healthy Americans: Our American Heart Association Diet* may be ordered from the Web site.

## National Heart, Lung, and Blood Institute
**<http://www.nhlbi.nih.gov>**

Select "Recipes for Healthy Eating" for a wide variety of heart-healthy recipes for many ethnic groups.

## Ornish Program

*See* **Preventive Medicine Research Institute.**

## Preventive Medicine Research Institute (PMRI)
**<http://www.pmri.org>**

PMRI is a nonprofit foundation conducting research on the effect of lifestyle changes on reversal of heart disease. Dean Ornish, MD, is the founder and director of this Sausalito, California, organization. Select "Lifestyle Advantage" to read about this collaborative effort between the Institute and Highmark. Lifestyle Advantage markets the Dr. Dean Ornish Program for Reversing Heart Disease, a commercial program available at approximately fifteen hospitals in Pennsylvania and West Virginia. A government-sponsored project is available for Medicare recipients: Medicare Lifestyle Modification Program Demonstration. The study wants to determine whether lifestyle change is a cost-effective alternative to traditional medical management aimed at reversing heart disease.

Dr. Ornish has a number of links set up on WebMD Health (http://my.webmd.com). The "Dean Ornish, MD's Lifestyle Program" provides information and support to people following his lifestyle guidelines.

## Pritikin Longevity Centers

*See* Chapter 10, "Weight Loss Spas and Residential Diet Programs."

# HEMORRHOIDS

*See also* HIGH-FIBER DIET (in this chapter).

## Milton S. Hershey Medical Center. Health & Disease Information. A to Z Topics. Hemorrhoids. <http://www.hmc.psu.edu/healthinfo>

This site offers good advice on the benefits of eating a high- fiber diet and drinking plenty of fluids. The site indicates which foods are high in fiber.

# HEPATITIS C

## Hepatitis C Council of South Australia. Diet and Hepatitis C. <http://www.hepccouncilsa.asn.au/diet.html>

This is a nonprofit support group that receives funding from the Department of Health of the government of South Australia. The diet section is very informative, but U.S. patients will need to convert metric measurements to U.S. equivalents.

# HIGH-CALCIUM DIET

## American Association of Clinical Endocrinologists (AACE). Calcium Content of Various Calcium-Rich Foods. <http://www.aace.com/pub/pdf/guidlines/osteoporosis2001 Revised.pdf>

Print page 10 (Table 5) of AACE's guidelines for the prevention and treatment of postmenopausal osteoporosis. Tape the list of foods to the refrigerator for monitoring daily intake of calcium-rich foods. Supplements provide additional calcium, but eating calcium-rich foods is a better way to meet daily requirements. The rest of the report is informative, too, especially for patients newly diagnosed with osteoporosis.

## Powerful Bones. Powerful Girls.
## <http://cdc.gov/powerfulbones>

This is part of the government initiative, National Bone Health Campaign, which is aimed at girls ages nine to twelve. Physicians and nutritionists now know that females need to start protecting their bones and maintaining calcium resources at an early age—not just at menopause. The "Toolbox" features information on calcium content of foods, a grocery list, and free materials such as place mats, stickers, and clipboards to encourage girls to get more physical exercise and consume calcium-rich foods (see Figure 8.3).

FIGURE 8.3. Powerful Bones
<http://www.cdc.gov/powerfulbones>

## HIGH-FIBER DIET

**Johns Hopkins Bayview Medical Center. Clinical Nutrition: High Fiber Diet**
**<http://www.jhbmc.jhu.edu/NUTRI/fiber.html>**

This site offers an excellent explanation of the difference between soluble and insoluble fiber. Print the list of fiber-rich foods.

## HIGH-POTASSIUM DIET

**University Of Maryland Medical Center. Complementary Medicine Program. Potassium**
**<http://www.umm.edu/altmed/ConsSupplements/Potassiumcs .html>**

Select "Dietary Sources." Ignore the section on potassium supplements unless your physician advises you to take supplements.

## HYPERTENSION (HIGH BLOOD PRESSURE)

**National Heart, Lung, and Blood Institute (NHLBI). DASH Eating Plan**
**<http://www.nhlbi.nih.gov/health/public/heart/hbp/dash>**

Dietary Approaches to Stop Hypertension, the DASH Diet, is recommended by NHLBI to reduce hypertension. Download the complete plan from this site. The diet is low in fat and rich in fruits, vegetables, and low-fat dairy products. Also read the material on the DASH Diet at MayoClinic.com (http://www.mayoclinic.com). This site contains a history of the development of the diet and discusses the low-sodium variant: DASH+ Sodium.

## HYPOALLERGENIC DIETS

*See* FOOD ALLERGIES (in this chapter).

## HYPOGLYCEMIA

### Hypoglycemia Support Foundation. What is a Hypoglycemia Diet?
### <http://www.hypoglycemia.org/diet.asp>

The medical director of this nonprofit organization is an osteopathic physician, Douglas M. Baird, DO. The site reprints an excerpt from *The Do's and Don'ts of Hypoglycemia,* written by the president and founder of the foundation, Roberta Ruggiero. Ms. Ruggiero is not a physician.

## JET LAG

### Anti-Jet-Lag Diet.com
### <http://www.antijetlagdiet.com>

This commercial resource is commissioned by the University of Chicago's Argonne National Laboratory. For a fee, the company will calculate a personalized diet based on your flight-departure city, date and time, and the arrival city, date, and time. The Argonne National Laboratory also takes into account the time breakfast is normally eaten.

## KETOGENIC DIET

The ketogenic diet was first formulated in the 1920s. It is an extremely low carbohydrate, high fat, and low protein diet. The diet is believed to force the body to burn fat, instead of glucose, for energy. Some physicians recommend the diet for children whose epileptic seizures fail to respond to conventional medications. The diet must be carefully calculated and monitored under the supervision of a dietitian and a physician. The ketogenic diet also restricts calories and liquid intake. Kidney stones are a potential side effect.

**Johns Hopkins Epilepsy Center. The Ketogenic Diet**
**<http://www.neuro.jhmi.edu/Epilepsy/keto.html>**

The center sells a computer disc explaining the diet to dietitians and parents, although it emphasizes that the ketogenic diet is a medical treatment that should be administered only under medical supervision.

**Stanford University. Lucile S. Packard Children's Hospital.**
**Seizures and Epilepsy. What Is a Ketogenic Diet?**
**<http://www.lpch.org/DiseaseHealthInfo/HealthLibrary/neuro/**
**seizep.html>**

This site thoroughly answers questions raised by parents about the diet. The hospital has been using the ketogenic diet since 1995.

## KIDNEY DISEASE/DIALYSIS

**National Kidney Foundation**
**<http://www.kidney.org>**

See the section "Living Well on Dialysis—a Cookbook for Patients and Their Families."

**Council on Renal Nutrition of the National Kidney Foundation**
**(NKF-CRN)**
**<http://www.kidney.org/professionals/CRN/nutrition.cfm>**

This site offers over one dozen specialized, patient-oriented brochures and fact sheets to order. A reference is offered to the National Renal Diet, developed by NKF-CRN and the American Dietetic Association (order from American Dietetic Association, 216 W. Jackson Blvd, Chicago, IL 60606-6995). It provides a link to "Patient Education Pages" in *The*

*Journal of Renal Nutrition* and a list of cookbooks for kidney patients. Each listing is annotated with comments on content and includes ordering information.

The site includes the full text of some free brochures: "Nutrition and Chronic Kidney Disease," "Nutrition and Transplantation," "Nutrition and Hemodialysis," and "Dining out With Confidence: a Guide for Patients With Kidney Disease."

## *LACTOSE-FREE DIET*

*See also* **National Dairy Council (NDC)** in Chapter 4, "General Diet and Nutrition Web Sites."

### Milk Makes Me Sick: Exploration of the Basis of Lactose Intolerance
**<http://www.exploratorium.edu/snacks/milk_makes-me_sick/index.html>**

This site includes a science lesson and experiment ("Science Snacks") written by Karen E. Kalumuck of the San Francisco Exploratorium: The Museum of Science, Art, and Human Perception.

### No Milk Page
**<http://www.panix.com/~nomilk>**

No Milk is a very extensive gateway to resources on all aspects of lactose intolerance and milk-free diets. It runs the gamut from MilkSucks .com to Web sites produced by government organizations and universities. Don Wiss maintains the Web site. MilkSucks.com is a site produced by PETA (People for the Ethical Treatment of Animals).

## *LEUKEMIA*

*See also* CANCER (in this chapter).

## Cancer Treatment Centers of America. Leukemia Cancer Center. Nutritional Therapy
<http://www.cancercenter.com/leukemia/nutritional-therapy.cfm>

Cancer Treatment Centers of America is a group of hospitals that uses "patient-empowered medicine" combining traditional medical, surgical, and radiological cancer treatment with spirituality and mind/body interventions. Read the advice on diet and nutritional supplements. View the link for nutritional recipe cards.

### *LOW-CARBOHYDRATE DIETS*

*See* Chapter 6, "Weight Loss (Nonsurgical) Web Sites," the entries for **Atkins Nutritionals, Carbohydrate Addict's Official Website, South Beach Diet Online,** and **The Zone Diet.**

### *LOW-CHOLESTEROL DIET*

## The Basics of a Low-Cholesterol, Low-Fat Diet (see also Atherosclerosis, Heart Disease, and High Cholesterol)
<http://cholesterol.about.com/cs/controlwithdiet/a/goodfood.htm>

This site contains good advice from Jennifer Moll, a food microbiologist and graduate student studying for a PharmD degree.

### *LOW-FAT DIET*

## Fatfree: The Low-Fat Vegetarian Recipe Archive
<http://www.fatfree.com>

On this site, search an archive of nearly 5,000 fat-free and very low-fat-free recipes.

## LOW-FIBER DIET

See **MayoClinic.com.**

## LOW-POTASSIUM DIET

**Clinical Nutrition: Low Potassium Diet**
**<http://www.jhbmc.jhu.edu/NUTRI/potassium.html>**

If a physician recommends a low-potassium diet (usually because of kidney problems), follow the guidelines at the Johns Hopkins Bayview Medical Center Web site.

## LOW-PURINE DIET

See GOUT.

## LOW-RESIDUE DIET

See also **MayoClinic.com.**

**University of Louisville Health Care. Digestive Health Center.**
**Patient Resource Center. Low Residue Diet**
**<http://www.uoflhealthcare.org/digestivehealth/lowresdiet.htm>**

This site has patient education material from the Digestive Health Department of University of Louisville Health Care, a major teaching hospital in Kentucky. Read the list of low-residue foods and the accompanying explanation of why low-fiber foods are easily digested, leaving little residue in the intestinal tract. Low-fiber diets are sometimes prescribed after gastrointestinal surgery.

## LOW-SODIUM DIET

### MedlinePlus Medical Encyclopedia: Diet for People with Chronic Kidney Disease
<http://medlineplus.gov>

On this site, select "Medical Encyclopedia," select "Kidney Disease—Diet," and link to "Diet for People with Chronic Kidney Disease."

## MALABSORPTION SYNDROME

*See* CELIAC DISEASE (in this chapter).

## MONOAMINE OXIDASE INHIBITOR (MAOI) DIET

*See also* HEADACHES/MIGRAINE ELIMINATION DIET (in this chapter).

### MayoClinic.com. MAOI.
<http://www.mayoclinic.com>

On this site, select "Food and Nutrition," then "Special Diets," and type "MAOI" at the search prompt. Patients who take monoamine oxidase inhibitors (MAOIs), usually to treat depression, must restrict the amount of tyramine in their diet. MayoClinic.com includes a very useful table of foods to avoid. Note the advice on avoiding leftovers and restaurant foods unless freshly prepared.

## MIGRAINE HEADACHES

*See* HEADACHES/MIGRAINE ELIMINATION DIET (in this chapter).

## *MSG-FREE DIET*

### Migraine Web: Tips on Starting an MSG-Free Diet
### \<http://www.migraineweb.com/index.html\>

This site is very informative, although there is no information about who created the Web site or his or her credentials! The Eating Out section discusses MSG-free (monosodium glutamate) menu items of several major restaurant chains.

## *MULTIPLE SCLEROSIS*

### National MS Society
### \<http://www.nationalmssociety.org\>

Select "Living with MS," then "Healthy Living with MS" to read advice about eating a healthy diet. This organization does not recommend fad diets that promise to cure multiple sclerosis.

## *OSTEOPOROSIS*

*See* HIGH-CALCIUM DIET (in this chapter).

## *PHENYLKETONURIA*

### University of Washington PKU Clinic
### \<http://depts.washington.edu/pku/diet.html\>

This site offers excellent material geared to parents, from the University of Washington in Seattle. Phenylketonuria (PKU) is a genetic disorder where the body does not manufacture enough of a particular liver enzyme. Patients must strictly adhere to a special diet.

## PREGNANCY, LACTATION, BREAST-FEEDING

Approximately 60 percent of women breast-feed their babies. In the 1950s the rate was 20 percent.

### La Leche League International (LLLI)
### <http://www.lalecheleague.org>

LLLI is a nonprofit organization providing information and advocacy and encouraging breast-feeding by means of woman-to-woman support. The fifty-year-old organization sponsors seminars for health care professionals and publishes materials for consumers and physicians. The site includes directories of accredited LLLI speakers and group meetings (see Figure 8.4).

FIGURE 8.4. La Leche League International
<http://www.lalecheleague.org>
Reprinted with permission of La Leche League International.

**March of Dimes: During Your Pregnancy: Your Healthy Diet**
**<http://www.marchofdimes.com/pnhec/159_823.asp>**

Many Web sites offer advice on nutrition for pregnant women, but some information is written by "experts" with questionable credentials. The March of Dimes' mission is to prevent birth defects. Adequate nutrition and healthy habits during (and before) pregnancy are essential. Read the sections "Eating for Two" and "Your Healthy Diet." There is a list of habits and foods to avoid and suggestions for daily servings of government-recommended food groups. Unlike some sites, the March of Dimes has kept current with the 2005 revised food pyramid guidelines and notes that the new guidelines do not yet address the needs of pregnant women. The March of Dimes will publicize the recommendations as soon as they are released by the federal government.

**Maternal Diet**
**<http://www.unu.edu/unupress/unupbooks/80338e/80338E00**
**.htm>**

This is the full text of a report from a 1981 conference held at Darwin College in Cambridge, England. It offers detailed scientific information on nutritional requirements of lactating mothers. Of particular interest is a table indicating various nutrient requirements for lactating and nonlactating women.

**National Women's Health Information Center. Breastfeeding**
**<http://www.4woman.gov/breastfeeding>**

This site is part of WomensHealth.gov., a federal government resource on women's health. A current advertising campaign encourages new mothers to breast-feed their babies (National Breastfeeding Awareness Campaign). The Web site includes Q&A, information on the benefits of breast-feeding, news, and sections on storage, pumping, and other issues related to family life and workplace effects on breast-feeding.

## PROSTATE DISORDERS

### Prostate Cancer Foundation. Lycopene for the Prevention of Prostate Cancer
**<http://www.prostatecancerfoundation.org/nutrition>**

This site includes a clear summary of research on lycopene consumption as it relates to prevention of prostate cancer. Lycopene is the substance found in some red-colored foods, such as tomatoes, watermelons, papayas, and pink grapefruit. Select "Nutrition and Lifestyle," then "Nutrition and Prostate Cancer." Read pages 12-15 of this guide.

## RAW-FOODS DIET

This is an extreme diet regimen, so please be sure to consult with your physician before embarking on this type of diet.

### The Diet Channel—Raw Food Diet
**<http://www.thedietchannel.com/Raw-food-diet.htm>**

The Diet Channel gives a brief overview of the diet. The site deals with two issues of concern: lack of protein and pesticides in nonorganic fruits and vegetables. Individuals on the raw-food diet will need to obtain enough protein by eating nuts and seeds. Organic produce is free of pesticides.

## REFLUX ESOPHAGITIS/GERD/HEARTBURN

### RealAge Heartburn Center. Acid Reflux Diet
**<http://www.heartburn.realage.com/content.aspx/topic/9>**

Print the list titled "Heartburn Foods That Can Aggravate Heartburn." The trigger foods include chocolate, spearmint, peppermint, fatty foods, and fried foods.

**University of Louisville Health Care. What Is Gastroesophageal Reflux Disease (GERD) Resource Center—Digestive Health Center**
**<http://www.uoflhealthcare.org/digestivehealth/gerd.htm>**

The University of Louisville Health Care is a major academic medical center and hospital in Kentucky. The section on GERD is written for consumers. The color illustration of the physiological process of GERD is excellent. Read the information about the popular myth that a bland diet is best for preventing heartburn. The site suggests, instead, that reflux sufferers keep a careful log of foods and behaviors that trigger symptoms. Tight clothing, for example, can trigger heartburn.

## SEIZURE DISORDERS

*See* KETOGENIC DIET (in this chapter).

## SPECIAL DIETS, MISCELLANEOUS

*See also* **Wholefoodsmarket.com** under ORGANIC DIETS in Chapter 9, "Web Sites for Diets Reflecting Religious or Philosophical Beliefs or Lifestyles."

**MayoClinic.Com. Food and Nutrition Center. Special Diets**
**<http://www.mayoclinic.com>**

This site offers an extensive amount of authoritative information on diets for digestive disorders, hypertension (high blood pressure), and mineral imbalances. Searching by keyword at the search prompt can retrieve additional special diets.

**National Listing of Fish Advisories. Fact Sheet**
**<http://www.epa.gov/waterscience/fish/advisories/factsheet.pdf>**

See page 5 of the fact sheet ("National Advice Concerning Mercury in Fish") for information directed at young children, pregnant women, women who may become pregnant, and nursing mothers. The fact sheet, while acknowledging the benefits of fish and shellfish, cautions these groups to avoid certain fish and shellfish that contain very high levels of mercury and to limit weekly consumption of canned light tuna, shrimp, salmon, pollock, and catfish to no more than 12 ounces. Because of higher mercury levels, these vulnerable consumers should eat no more than 6 ounces a week of albacore tuna.

### SUBSTANCE ABUSE (ALCOHOLISM, ILLICIT DRUGS, PRESCRIPTION DRUGS)

**University of Maryland Medical Center. Diet and Substance Abuse Recovery**
**<http://www.umm.edu/ency/article/002149.htm>**

Read the thorough discussion of various types of substance abuse and addictions and how each affects the body nutritionally. Prevention of malnutrition, in general, and deficiency in specific nutrients is clearly explained. Some patients develop food addictions or substitute caffeine or sugar as they recover from drug addictions. The article recommends a low-fat, high-fiber, high-protein diet.

### TYRAMINE-FREE DIET

*See* MAOI DIET (in this chapter).

### WHEAT-FREE DIET

*See* CELIAC DISEASE (in this chapter).

### NOTE

1. Cameron, B.D. (2002) Celiac Disease: Internet Resources. *Health Care on the Internet* 6: 23-32.

Chapter 9

# Web Sites for Diets Reflecting Religious or Philosophical Beliefs or Lifestyles

What is food to one, is to others bitter poison.

Lucretius

## *HALAL*

In the halal diet, foods are classed into three categories:

1. *Halal foods* are permitted foods, neither haram or mushbooh.
2. *Haram foods* are forbidden: alcohol, lard, and carnivorous animals, including birds. Most followers of the halal diet also prefer not to eat fish.
3. *Mushbooh foods* are those forbidden because it may be impossible to determine the origin of the ingredients or the ingredients are known to be derived from animals.

### Food in the Arab World
### <http://www.al-bab.com/arab/food.htm>

This site includes recipes from around the world, categorized by country or region (Moroccan, Lebanese, Palestinian, Middle Eastern).

*Internet Guide to Medical Diets and Nutrition*
© 2006 by The Haworth Press, Inc. All rights reserved.
doi:10.1300/5852_09

Includes a guide to spices. Emphasis is on vegetarian dishes. Note the link to the Islamic Food and Nutrition Council of America "What is Halal?" The section clearly explains halal, haram, and mushbooh (see Figure 9.1).

## Halal and Healthy
## <http://www.soundvision.com/Info/halalhealthy/>

This site offers intelligent halal diet advice, emphasizing multi-grains, fruits, and vegetables. Also read the section "Good Nutrition: Guidelines for Healthy Eating."

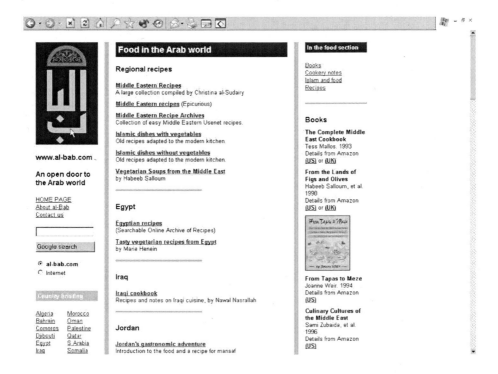

FIGURE 9.1. Food in the Arab World
<http://www.al-bab.com/arab/food.htm>
Reprinted with permission of Al-bab.com.

## Halal Definition
## <http://www.nutrition.co.th/halal.html>

Very thorough explanation of which foods can be consumed and detailed description of storage and preparation processes. Good reference for consumers and businesses.

## Muslim Consumer Group for Food Products
## <http://www.muslimconsumergroup.com>

The nonprofit and nonpolitical Muslim Consumer Group for Food Products was founded in 1993. Its aim is to educate Muslims about halal products in supermarkets and to educate food companies about requirements for receiving halal certification. Its halal symbol now appears on food products internationally. The Web site features a list of food ingredients categorized as "halal" (permitted), "haram" (forbidden), and "mushbooh" (suspect), based on Islamic dietary law. Haram foods are primarily those containing pork, alcohol, or human (usually hair) ingredients. The "News/ Alerts" section warns about brands of products, including cosmetics, containing haram (forbidden) ingredients. Particularly useful is a list of halal (approved) supermarket ingredients, arranged by product category. This list is kept current with a separate list "New Food Products." The section "Halal Food Selection" gives detailed advice, including how to determine whether an item with a kosher symbol meets halal requirements. A Q&A section is overseen by the Web site's senior food scientist, Syed Rasheeduddin Ahmed, who has experience in the food industry in research and development and quality assurance.


### *HINDU*

Under the Hindu diet code, all food is divided into three categories:

1. *Rajasic foods:* food derived from the meat of animals. Note that it is forbidden to eat beef or pork. Rajasic foods are supposed to produce activity and strong emotions. These foods contain onions and spices.

2. *Sattvic foods:* foods that do not irritate the digestive tract. Most Sattvic foods are fruits, nuts, and vegetables. Sattvic foods are supposed to promote tranquility and noble nature.
3. *Tamasic foods:* considered to be the worst type of food. Tamasic foods are leftover or spoiled food and are thought to produce jealous or greedy behavior.

## Hindu Universe—Health and Lifestyle
## <http://www.hindunet.org/healthlifestyle/>

Hindu Universe is a portal, containing links to news, discussion groups, lifestyle, and advertisements. It is one of the Web sites maintained by Hindunet: Global Hindu Electronic Networks, a project of the Hindu Students Council, based in Boston. Select "Vegetarianism," then select "More . . . ," You are now on the home page of Bhojan.org. Select "Food in Hindu Dharma," then "What is the Hindu Diet Code?" This section describes Rajasic foods (meats, other than beef or pork) and Sattvic foods (fruits, nuts, and vegetables). Generally, the Hindu diet excludes meat and dairy products. Eggs and (sometimes) seafood are permitted.

Also at the Bhojan.org Web site see "Resources for Vegetarians," which includes advice on buying fruits and vegetables, lists of vegetarian restaurants and grocery stores in the United States, and a detailed description of special airline diets (diabetic, gluten-free, low-purine, nonlactose, low-fat, low-sodium, lacto-ovo, kosher, Muslim, and Hindu) and whether they are suitable for vegetarians. The site includes an index of Hindu vegetarian recipes.

## *HOLISTIC NUTRITION*

## Institute for Integrative Nutrition
## <http://www.integrativenutrition.com>

The Institute for Integrative Nutrition, accredited by the American Association of Drugless Practitioners, trains health counselors. The students study dozens of diets and integrate theories to build recipes based on "primary" foods, such as whole grains, vegetables, and fish. Select "Resources," then "Recipes" to learn how to prepare a variety of healthy dishes.

## *KOSHER*

The following lines from Leviticus (11:3; 11:7-8) summarize the kosher guidelines:

- Whatsoever parteth the hoof, and is cloven-footed, and cheweth the cud, among the beasts, that ye shall not eat.
- And the swine . . . is unclean to you. Of their flesh shall ye not eat.

### Kashrut.com
### <http://www.kashrut.com>

This Web site dubs itself "The premier kosher information source on the Internet." Included on this site are many kosher recipes, including breads and beverages, and an annotated list of cookbooks that can be purchased online (some of the profits help fund Kashrut.com). See "Product Updates" for newly certified food products, bakeries, and restaurants.

### Kosher Delight
### <http://kosherdelight.com>

International in scope, Kosher Delight is a portal to a wide variety of resources. Search "The Best Kosher Restaurant Guide in the World" to see listings of kosher food establishments (more markets and caterers than restaurants). The site also has a kosher recipe section.

### Machers.com
### <http://www.machers.com>

This Jewish portal ("Your gateway to the Jewish community") pulls together several useful educational resources. Select "Kosher Guide" from the index on the left side of the home page and then "Laws and

Guidelines" and then "Kashrut Guide." This links to "What is Kosher" by Rabbi Y. Baumgarten of South Africa. The site is sponsored by the Union of Orthodox Synagogues of South Africa. Rabbi Baumgarten is Principal Rabbinical Supervisor of the Kashrut Department of the Johannesburg Beth Din. He explains, in lay terms, the basic principles of kashrut (kosher dietary laws), the separation of dairy, meat, fish, and pareve (neutral) foods. "Kosher" is the Hebrew word meaning "fit." Kosher foods and beverages are "fit to be ingested by one who adheres to the laws of kashrut." Detailed sections are included on identifying kosher species of animals and their ritual slaughter, kosher butchery, and setting up the kosher kitchen. The section on the kosher kitchen describes the separation of meat from milk and the requirement to own two separate sets of cookware, cutlery, and dishes. This site could serve as a primer for new converts to Judaism or for anyone else needing instruction in the details of observing kosher dietary laws. The rabbi explains that the information is geared to consumers, and he includes the Johannesburg Beth Din's e-mail address and telephone number for readers who have questions.

Under "Laws and Guidelines," see the link to "Kosher Consumers Union," an independent consumer-advocacy group based in the United States. Its goals are similar to that of other consumer- advocacy organizations: recommendations, recalls, and warnings on products—kosher foods and products in this instance. Observant Jews will find the section "Kosher Alerts" to be invaluable, since it alerts readers to contaminated foods and foods incorrectly labeled "kosher." The "Recommended Food List" features specific brands of food items as recommended by the Kosher Consumers Union of Lakewood, New Jersey. The "Mini Kashrus [Kashrut] Course" is more succinct than the previously mentioned kosher dietary-law Web sites.

Another link on Machers.com is Judaism 101's "Kashrut: Jewish Dietary Laws," by Tracey R. Rich, Web master at JewFAQ.org. This is also a great reference resource, but it reads more like a college text, complete with "hot" links to the Torah and other references. It also includes a chapter on the reasons a person might want to be "observant" (of kosher dietary laws).

Another section of Judaism 101 is "Jewish Cooking." This section gives recipes for a variety of kosher dishes. Befitting the textbook arrangement, this portion also includes background information on Jewish cuisine.

"Kosher on Campus," written by the United Synagogue Community Development Department, is geared to Great Britain, but any Jewish student would find the advice useful. Sections include "Kashrut Basics," "Making Your Kitchen Kosher," and "What Foods are Kosher?"

## MACROBIOTIC

The term *macrobiotic* comes from the Greek words *macro* (long) and *bios* (life). The macrobiotic diet consists mainly of whole grains, cereals, and organic, locally grown, cooked vegetables. Soups based on seaweed and fermented soy (miso) constitute a small part of the diet. More than a diet, it is a way of life combining Buddhism with a vegetarian diet. A Japanese writer and philosopher, George Ohsawa, sought to combine some Christian teachings with Zen Buddhism and portions of ayurvedic (Asian) and Western medicine. Ohsawa came to the United States in the 1960s, and a disciple, Michio Kushi, opened the Kushi Institute in Boston in the late 1970s. The American Cancer Society refutes proponents' claims that the diet can prevent cancer, although scientific research is underway. The National Institutes of Health Office of Alternative Medicine is funding a pilot study to determine whether a macrobiotic diet can prevent cancer. Some variations of the macrobiotic diet are so restrictive as to be considered dangerous, so always consult with a registered dietitian or physician before undertaking a radical diet change. For a thorough description of the history of macrobiotic diet and philosophy and possible dangers of the diet, read the material at the American Cancer Society's Web page (http://www.acs.org). The section on the macrobiotic diet appears within the section "Making Treatment Decisions: Diet and Nutrition." In addition to the Web sites listed here, be sure to read the section on macrobiotic diets on the Whole Food Market Web site (see this site's listing under Organic Diets in this chapter).

## Christina Cooks
## <http://www.christinacooks.com>

This is the Web site of Christina Pirello, host of the popular PBS television series *Christina Cooks.* When Pirello was twenty-six, she was diagnosed with advanced leukemia. After her future husband, Robert, gave her a book by Michio Kushi, *The Cancer Prevention Diet,* she changed her lifestyle and way of eating. Nearly two decades later, there is no sign of the leukemia. She attributes her good health to the adoption of a macrobiotic diet.

Her Web site features recipes from her show. She advertises merchandise, trips, and books, so this is a decidedly commercial site, but the diet she advocates is not extreme, emphasizing whole foods, grains, and vegetables.

## CyberMacro: Macrobiotics and Macrobiotic Foods Home
## <http://www.cybermacro.com>

Gary Miller's Web site features recipes, news, articles, macrobiotic health forums, a glossary of food-related definitions, an online catalog of macrobiotic foods, Japanese cookware, books, and links to other macrobiotic sites. Some sections require registration, but it is free.

## Healing Cuisine.com
## <http://www.healingcuisine.com>

Meredith McCarty, a certified diet counselor and nutrition educator, created Healing Cuisine in 1995. A food coach, she offers individual consultations, lectures, writes books, and leads tours of farmers' markets and natural food stores. The recipes consist of vegetarian, natural food, and macrobiotic dishes.

## Kushi Institute
## <http://www.kushiinstitute.org>

Michio Kushi founded this educational institute in 1978. Today it is located in Becket, Massachusetts. Mr. Kushi wrote *The Cancer Prevention Diet, The Macrobiotic Way, The Book of Macrobiotics,* and other books. The institute offers residential educational programs. The site features personal stories of healing (supposedly due to the macrobiotic diet and lifestyle), a newsletter, discussion group, online store, and recipes. Choose "Healing" and read the section "What is Macrobiotics?"

## Macrobiotic Diet Basics
## &lt;http://www.macrobioticcooking.com&gt;

This site offers recipes and an online tutorial with the seven essential components of the macrobiotic diet. Macrobiotic Diet Basics is the Web site of Chef Linda Wemhoff.

## South River Miso
## &lt;http://www.southrivermiso.com&gt;

Miso is a fermented food usually made from soybeans, cultured grain, and sea salt. It is usually used as a high-protein seasoning in soup. South River Miso sells several types of miso, organic brown rice, maple syrup, sesame tahini, and other food products. The Web site has recipes and information about the history and process of making miso.

## *ORGANIC DIETS*

Organic foods are grown without pesticides and prepared or preserved without artificial preservatives. Organic foods are never genetically modified. Animals used for meat or milk are never injected with antibiotics or growth hormones. Organic vegetables and fruits are said to be more nutritious and flavorful.

## Natural Health: Building Blocks for Your Body
## &lt;http://www.naturalhealth.com&gt;

Natural Health's philosophy for a healthy, nutritional life is balanced: "Organic when practical, natural when possible, plant-based when available, and taste and appearance as good as you can make it with the time available." It recognizes that supplements are still usually necessary for good nutrition. The site includes a directory of online grocers, articles on herbal remedies, and links to natural health and fitness Web sites.

## Organic Consumers Association (OCA)
## <http://www.organicconsumers.org>

OCA is a grassroots, nonprofit, public-interest organization that advocates for safe food, fair trade, organic farming, environmental sustainability, corporate accountability, and an end to genetically engineered food. Select the tab "Find Organics" to link to GreenPeople, a directory of organic foods, baby and beauty products, and recycled products. The Web site also contains a chat room, news stories, and alerts ("Mad Cow Disease"). Read their online newsletter *Organic View.*

## Trader Joe's
## <http://www.traderjoes.com>

Trader Joe's is a chain of over 200 stores across the United States. Although it does not sell, exclusively, organic products, (approximately 50 percent of its products are organic), the site contains a concise explanation of how certain foods comply with the United States Department of Agriculture's National Organic Program (http://www.usda.ams.gov/nop). It explains the difference between a product describing itself as "organic" versus "made with organic ingredients." Select the section "Choosing and Using Our Products" and click on "Organic Products." Trader Joe's site lists the organic products carried in its stores according to geographic region (West Coast or East Coast/Midwest).

## Whole Foods Market
## <http://www.wholefoodsmarket.com>

Whole Foods Market is the largest chain of markets selling natural and organic foods. It has 160 stores in the United States and Great Britain. The company supports local food banks and contributes a portion of its profits to nonprofit groups. Whole Foods Market supports organic farming as a way to promote sustainable agriculture, protect the environment, and support farmworkers.

The Web site has a section "Health Info" which contains links to reputable sites such as *HerbalGram* and *HerbClip,* online publications of the American Botanical Council. "Health and Wellness Topics" should be bookmarked for your personal online reference library. Within this section, select "Health & Wellness Topics" to view excellent information on safe handling of food, reading food labels, advice on nutrition for mothers and children, and extensive information on special diets. The information on the history of the macrobiotic movement is especially thorough. The section on gluten-free diets includes recipes.

## VEGETARIAN DIETS (ALL VARIATIONS)

Over 12 million Americans follow a vegetarian diet for ethical, religious, or health reasons. Generally speaking, a vegetarian diet is one that excludes animal products. Some vegetarians, however, consume eggs, honey, milk, or fish. The following Web sites provide practical, ethical, philosophical, social, and political information and support for the many vegetarian communities.

**Eat Veg.com**
**<http://www.eatveg.com>**

On this site you can read advice on how to transition to vegetarianism. It offers many ads and services ("VeggieDate" personals, juicers, dehydrators), but there is a lot of good information on this site. "Why a Plant-Based Diet," for instance, makes health claims for swearing off meat.

**HappyCow's Vegetarian Guide to Restaurants and Health Food Stores**
**<http://www.happycow.net>**

HappyCow has a very extensive worldwide directory of places to eat and stores to shop. Readers are encouraged to submit their own restaurant reviews. The Web site also has book reviews (including cookbooks), veggie humor, travel information, a vegetarian dating board, and animal rights information.

## North American Vegetarian Society (NAVS)
## <http://www.navs-online.org>

NAVS is a nonprofit educational organization. The Web site includes recipes and links to sites listing vegetarian restaurants around the world. NAVS sells cookbooks, gardening books, mugs, T-shirts, and other items to support the organization.

## VegCooking
## <http://www.vegcooking.com>

VegCooking is produced by PETA, People for the Ethical Treatment of Animals. The site is very attractive, with large photographs accompanying recipes and articles featuring school lunches, celebrity galas, vegetarian camping, and profiles of vegetarian chefs. VegCooking reads just like your favorite lifestyle magazines, but with a vegetarian slant.

## Vegetarian Resource Group (VRG)
## <http://www.vrg.org>

VRG is well organized and gives balanced information without stridency. Nutrition myths and concerns are addressed with intelligent, well-documented material. Read information about calcium, iron, and vitamin $B_{12}$ requirements. The recipes at this site are creative. Readers can also download selected recipes from the magazine *Vegetarian Journal*. VRG compiles free dining guides to major U.S. cities. Read "Choosing and Using a Dietitian" and nutrition guides for seniors, children, and pregnant women (see Figure 9.2).

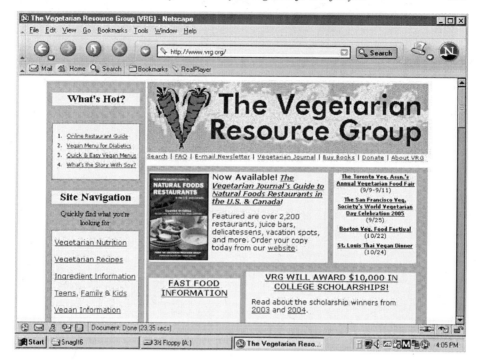

FIGURE 9.2. Vegetarian Resource Group
<http://vrg.org>
Reprinted with permission of the Vegetarian Resource Group.

## Vegetarian Source Online
## <http://www.VegSource.com>

VegSource readers can participate in over thirty discussion forums. There are sections on weight loss, parenting, vegetarian FAQs, and news items.

## *Vegetarian Times*
## <http://www.vegetariantimes.com>

This online magazine includes recipes that cover the gamut of vegetarian styles: vegan (no dairy or eggs, may contain honey), lacto (dairy but no eggs), and ovo-lacto (contains eggs and dairy). Full nutritional information is provided with each recipe. The editorial advisory board is composed of physicians (Drs. Dean Ornish and Neal Barnard), dietians, athletes, and a celebrity author (*Moosewood Cookbook* writer Mollie Katzen).

## VegWeb
## <http://www.vegweb.com>

The VegWeb site has a more traditional vegetarian approach. Political/ethical issues such as animal rights are downplayed. There are articles on composting and over thirty categories of recipes. A useful feature is a section reviewing different types of vegetarian cookbooks.

Chapter 10

# Weight Loss Spas
# and Residential Diet Programs

## Canyon Ranch Health Resorts
## &lt;http://www.canyonranch.com&gt;

Mel Zuckerman founded Canyon Ranch in 1979. Zuckerman was influenced by holistic medicine practitioner Dr. Jesse F. Williams, MD, whose concept of health was not merely the absence of illness, but a condition that makes it possible for one to experience the highest enjoyment of life. This is the philosophy of Canyon Ranch Health Resorts and Canyon Ranch SpaClubs. Today there are Canyon Ranch Health Resorts in Tucson, Arizona, and Lenox, Massachusetts. The medical staff at both facilities consists of board-certified physicians, dietitians, physical therapists, exercise physiologists, behavioral health specialists, acupuncturists, and other specialists. The Web site publishes brief biographies for all of these clinicians.

Guests can choose to focus on one of several health programs, including "Ultrametabolism: New Approaches to Weight Loss" and "Ultraprevention," which focuses on nutrition, fitness, and strategies to enhance longevity and energy. Guests attend hands-on cooking classes to learn healthy eating.

## The Cleanse
## &lt;http://www.thecleanse.com&gt;

Kartar Singh Khalsa, a doctor of oriental medicine, acupuncturist, and Kundalini yoga practitioner is the director of The Cleanse Vegan detoxification program. The residential program in northern New Mex-

*Internet Guide to Medical Diets and Nutrition*
© 2006 by The Haworth Press, Inc. All rights reserved.
doi:10.1300/5852_10

ico aims to detoxify and cleanse the inner body. It features Kundalini yoga exercises, Chinese and Ayurvedic herbs, juices, vitamins, enemas, and a basic alkaline diet (no dairy, meat, wheat, alcohol, caffeine, or refined sugar).

## Miraval Life in Balance Resort
## <http://www.miravalresort.com>

Miraval is located in Tucson, Arizona. Center for Life in Balance programming includes a weight-loss program: "Weight in Balance." Guests wishing to concentrate on nutrition or weight loss can attend cooking demonstrations, dine with a nutritionist, and attend classes on food labels, mindful eating, weight loss without dieting, food pantry reorganization, and mindful dining out.

## Dr. Dean Ornish Program for Reversing Heart Disease
##   at Windber Medical Center

*See also* discussion of Ornish Program under ATHEROSCLEROSIS OR HIGH CHOLESTEROL in Chapter 8, "Diet, Nutrition, and Recipe Web Sites for Medical Problems and Conditions."

This is one of several residential programs (one-day, three-day, and weeklong retreats) sanctioned by Dr. Dean Ornish. It is recommended for patients contemplating coronary bypass surgery or who have experienced a cardiac event, have coronary artery disease, or significant high blood pressure, high cholesterol, or obesity. Windber Medical Center is a hospital in Somerset County, Pennsylvania. The program features lifestyle changes that include a low-fat, whole-grain, vegetarian diet, moderate exercise, and stress reduction techniques. A registered dietitian manages the program.

## Pritikin Longevity Center and Spa
## <http://www.pritikin.com>

Nathan Pritikin, Dr. David Lehr, and Dr. Robert Bauer founded the Pritikin Longevity Center in 1978. Today, Paul Bauer, the son of Robert Bauer, directs the program, in Aventura, Florida. The medically supervised program consists of exercise and a diet that is low fat, low salt, and low in refined carbohydrates. The diet is high in fiber, antioxidants, phytochemicals, vitamins, and minerals and permits modest amounts of fish and low-fat dairy products. The program is successful in aiding weight loss, relieving the pain of arthritis, and controlling high cholesterol, athero-sclerosis, and diabetes.

## Rice Diet Program
**<http://www.ricedietprogram.com>**

The Rice Diet Program was developed in 1939 at Duke University in Durham, North Carolina. The residential "lifestyle" program special-izes in treating patients with serious chronic illnesses such as hyperten-sion, diabetes, kidney disease, heart disease, and obesity. Patients at-tend classes every day and are closely monitored by medical personnel. The Web site offers detailed information about the program, including fees, and descriptions of the phases of the low-fat, low-salt diet. The diet emphasizes fruit and grains, gradually adding vegetables and fish. Most patients stay for four to eight weeks.

## SpaFinder
**<http://www.spafinders.com>**

SpaFinder, a spa travel and marketing company, began in 1986. It publishes a book and a magazine in addition to the Web site. Select "Spas by Category" or "Medical Spa Program." SpaFinder covers in-ternational destination spas and day spas. Search by price, geographic location, and many other categories.

## We Care Spa
**<http://www.wecarespa.com>**

This holistic health spa, located in Desert Hot Springs, California, features fasting, an all-liquid diet, yoga, meditation, colonics, and mas-sages. The owner, Susana Lombardi, developed the stress-relieving, detoxification program herself through reading books and experienc-ing beneficial treatment by a holistic chiropractor.

Chapter 11

# Recipes and Discussion Group Web Sites

There is no accounting for tastes.

Anonymous Latin proverb

Also see recipe sections in many of the general nutrition and diet program Web sites.

## All Recipes
## <http://www.allrecipes.com>

This site's slogan is "Real recipes from real people." Search "All Recipes Collection" for diabetic, low-carb, low-fat, and vegetarian recipes. Find recipes for management of high blood pressure, high cholesterol, and general wellness in the section "Healthy Solutions." A year of customized meal planners ("Nutri-Planners") is available for purchase: Diabetes Management Program, Healthy Heart Program, or General Wellness Program.

## Cook's Thesaurus
## <http://www.foodsubs.com>

This site provides substitutions for recipe ingredients (ethnic substitutions, low fat, low cost, and low calorie). This cooking encyclopedia also covers cooking tools.

*Internet Guide to Medical Diets and Nutrition*
© 2006 by The Haworth Press, Inc. All rights reserved.
doi:10.1300/5852_11

## Cooking Light
## <http://www.cookinglight.com>

This is the Web site of the magazine with the same name. It has recipes plus "Cooking 101," with pantry ingredient lists, a food storage guide, and annual recipe index.

## Eat-L Foodlore/Recipe Exchange

This is a moderated discussion group. To subscribe, send an e-mail to listserv@listserv.vt.edu. Put nothing in the subject line, and type "subscribe Eat-L your name" in the body of the message (without quotation marks).

## Epicurious
## <http://www.epicurious.com>

Calling itself "the world's greatest recipe collection," Epicurious covers recipes, cooking, restaurants, and wine (see Figure 11.1).

## Flora's Recipe Hideout
## <http://www.floras-hideout.com/recipes/>

Flora has a dozen categories, with emphasis on desserts. The low-carb and low-fat categories are excellent and varied.

## Meals for You
## <http://www.mealsforyou.com>

This site allows you to search for recipes meeting common nutritional requirements: low fat, low calorie, low sodium, diet points (a la Weight Watchers), and low carb, etc.

FIGURE 11.1. Epicurious
<http://www.epicurious.com>
Reprinted with permission of Epicurious.com.

## Recipe Archive Index
### <http://www.cs.cmu.edu/~mjw/recipes>

The Carnegie Mellon University School of Computer Science currently hosts this site. The resource provides links to hundreds of recipes. Under "Greek dishes," there were six recipes for stuffed grape leaves!

## Recipe Corner
### <http://www.aicr.org>

The American Institute for Cancer Research produces this site. Select "Diet and Cancer" then "Recipe Corner." Search recipes by category or type in a particular ingredient. You can sign up to receive new healthy recipes each week.

## The Recipe Link
## <http://www.recipelink.com>

Recipe Link includes recipes from newspaper columns, recipes using brand-name ingredients, kosher recipes, romantic meal recipes, brunch recipes, and recipes that use particular ingredients.

## SOAR: Searchable Online Archive of Recipes
## <http://www.recipesource.com>

Recipe Source hosts SOAR, a very extensive grouping of recipes. It is particularly strong in ethnic cuisine.

Chapter 12

# Careers in Nutrition and Dietetics

Many colleges and universities offer bachelor's, master's, and doctoral degrees in nutrition and dietetics. A more recent occupation is nutritionist. Several educational programs for nutritionists are offered on the Internet, but few meet mainstream criteria for credentialing. Judge for yourself.

## American Dietetic Association
## <http://www.eatright.org>

Select "Careers and Students" to see information on various careers in dietetics, scholarships and financial aid, accredited programs, and education pathway flowcharts.

## Clayton College of Natural Health
## <http://www.ccnh.edu>

Clayton College of Natural Health offers bachelor's, master's and doctoral degrees in holistic nutrition and related areas. The school is twenty-five years old. The programs are offered online. The college is accredited by the American Association of Drugless Practitioners and the American Naturopathic Medical Accreditation Board.

## Institute for Integrative Nutrition
## <http://www.integrativenutrition.com>

This New York City school says it is the only school in the world integrating several major dietary theories, including blood-type diets,

*Internet Guide to Medical Diets and Nutrition*
© 2006 by The Haworth Press, Inc. All rights reserved.
doi:10.1300/5852_12

Atkins, ayurveda, macrobiotics, raw foods, the Zone, and the USDA MyPyramid. Holistic nutrition integrates all the different dietary theories. The institute believes there is no "perfect way of eating that works for everybody." Students are taught to recommend a diet after considering a patient's career, physical activity level, spirituality, and personal relationships. Graduates of the program call themselves "health counselors."

Chapter 13

# Happy Endings:
# Eat Something—You'll Feel Better!

Let us eat and drink; for tomorrow we shall die.

Isaiah 22:13

By now you probably realize that food can be a pleasure or a problem to different people. Some people overeat to the point that food is their drug. For others, certain diets can be therapeutic, even lifesaving. Cooking can be a hobby or a way of expressing love for our friends and family. Nutrition or cooking can be a profession. Food can be an extension of our religious beliefs or philosophy of living.

Remember that the Internet is dynamic: Web sites appear, disappear, or change address or arrangement. At the time this book was published, all Web site addresses were current. In most cases, if a Web site moves to a different location, there will be a "forwarding address," or the old site will automatically link to the new one. If this does not happen, try using a search engine to search for the name of the Web site. Sometimes a site is rearranged, and topics will appear on different "pages" of the Web site. Using the site's index should point you to the section you seek.

Do not forget to bookmark your favorite sites. You may find additional Web resources on the topics in this book, so add those to your folder of favorites.

*Internet Guide to Medical Diets and Nutrition*
© 2006 by The Haworth Press, Inc. All rights reserved.
doi:10.1300/5852_13

## AboutPizza.com
## <http://aboutpizza.com>

This Web site is produced by Omni Visions, a digital media company. The site pulls together links to the history of pizza, recipes, styles of pizza, popular ingredients, and news items from wire services ("Pizza truck stolen in Kalamazoo").

## Ben & Jerry's Homemade Ice Cream
## <http://www.benjerry.com>

On this site, get up close and personal with a delightful black and white cow! Select "From Cow to Cone" to see a humorous, yet informative animated narrative on the process of making ice cream.

## Chocophile.com
## <http://www.chocophile.com>

"Living the chocolate life" is the mantra of Clay Gordon, educator, writer, expert in high-definition television, computer graphics, and fine-chocolate lover extraordinaire! Gordon has pulled together a broad assortment of sites on everything chocolate: specialty chocolates (kosher, organic, vegan, and sugar-free), manufacturers (prestige, gourmet, mass market, baking chocolate, hot chocolate, and gourmet bars), styles (American, European, and French), and discussions and product reviews.

## TopSecretRecipes
## <http://www.topsecretrecipes.com>

Read through this site to learn how to make Mrs. Fields' chocolate chip cookies, IHOP pancakes, or Olive Garden pasta Alfredo. The recipes may not be authentic, but they are described as clones. View portions of each recipe before purchasing the detailed "kitchen clone." The author, Todd Wilbur, has written several books on brand-name recipes.

# Glossary

**activities calculator:** Enter current weight and desired number of calories to burn. Calculator suggests physical activities and how long it will take to burn off desired number of calories.

**bariatric surgery:** Weight-loss surgery.

**BMI:** Body mass index. A formula for standardizing the extent of obesity, based on body measurements. The formula: take weight in pounds divided by height in inches, squared, and then multiply that number by 703. A BMI between 19 and 24 is considered normal weight.

**BMR:** Basal metabolic rate. Amount of calories the body burns when at rest, but awake, over the course of one day. *See also* CALORIE/ENERGY NEEDS.

**calorie/energy needs:** Amount of calories a person needs to eat each day to maintain weight and fuel physical activity.

**calories burned:** Select type and duration of activity; calculator figures how many calories have been burned.

**clinically severe obesity:** Newer term for morbid obesity.

**dumping syndrome:** Group of unpleasant symptoms that may occur after weight-loss surgery due to the altered ability to digest sugars and fats and rapid movement of food from stomach to small intestines.

**duodenum:** First portion of the small intestine.

**glycemic index (GI):** A measurement of the impact a food has on blood-sugar and insulin levels. Researchers at the University of Sydney developed the index. The GI is the relative rate of blood sugar rising when a given food is compared with glucose, which has a GI of 100. Low-GI foods are thought to promote weight loss by providing a feeling of fullness (satiety) after a meal. Low-GI foods are thought to promote fat burning.

*Internet Guide to Medical Diets and Nutrition*
© 2006 by The Haworth Press, Inc. All rights reserved.
doi:10.1300/5852_14

Basmati rice has a low GI, while a baked potato has a high GI. Al dente spaghetti has a lower GI than spaghetti that has been boiled for 10 minutes. "Good carbs" have lower GIs than "bad carbs."

**halal:** Islamic term for "permissible." The set of Islamic dietary laws governing food processing, preparation, transport, storage, and consumption.

**jejunum:** Second portion of the small intestine.

**ketosis:** Process in which ketones are formed in the body during fasting.

**kosher:** Hebrew term for "fit" or "correct." Ritually fit for use in accordance with Orthodox Jewish laws or conforming to Jewish dietary laws.

**Kundalini yoga:** Philosophy of awakening potential energy and consciousness of the body and mind.

**macrobiotic:** A very restrictive diet consisting primarily of whole grains and unprocessed seaweed, nuts, and fermented foods such as pickles and sauerkraut. Consumption of small portions of fish is permitted on the diet.

**morbid obesity:** Condition where the BMI is 40 or greater. *See also* CLINICALLY SEVERE OBESITY.

**obesity:** Condition where the BMI is 30-39.9.

**overweight:** Condition where the BMI is 25-29.9.

**phytochemicals:** Natural bioactive compounds found in vegetables and fruits that work together with fiber and nutrients to protect against disease.

**Roux-en-Y gastric bypass:** Bariatric surgical procedure that combines restriction and malabsorption. Stapling or banding creates a small stomach pouch. A Y-shaped portion of the small intestine is attached to the pouch, bypassing the duodenum and a portion of the jejunum.

**tyramine:** Possible trigger of migraine headaches. Chemical found in ripened cheeses, herring, chocolate, pickled foods, sour cream, yogurt, nuts, onions, MSG, citrus foods, bananas, raisins, chicken livers, avocado, and alcoholic beverages.

**vegan:** A lifestyle that excludes the eating of any animal products or the wearing of fur, suede, or leather.

**vegetarian:** A person or diet that avoids animal products. Some vegetarians eat eggs, honey, and dairy products.

**vertical banded gastroplexy (VBG):** Bariatric surgical procedure that restricts the size of the stomach. A band and stapling create a small stomach pouch.

**waist-to-hip ratio:** Calculation used to determine whether body is pear- or apple-shaped.

# Further Reading

American Heart Association. (2005). *The no-fad diet: A personal plan for healthy weight loss*. New York: Clarkson Potter Publishers.

Blackwood, Hilary S. (2005). Help your patient downsize with bariatric surgery. *Med/Surg Insider* Fall: 4-9.

Duyff, Roberta Larson. (2002). *American Dietetic Association complete food and nutrition guide,* Second edition. Hoboken, NJ: John Wiley and Sons.

Dyer, Diana. (2002). *A dietitian's cancer story*. Ann Arbor, MI: Swan Press.

Hark, Lisa and Deen, Darwin. (2005). *Nutrition for life*. New York: Dorling Kindersley.

Havala, Suzanne. (2001). *Vegetarian cooking for dummies*. New York: Hungry Minds.

Hunter, Fiona and Gow, Emma-Lee. (2003). *Great healthy food for strong bones*, First edition. Toronto: Firefly Books.

Katz, David L. and Gonzalez, Maura Harrigan. (2002). *The way to eat*. Napierville, IL: Sourcebooks.

Khalsa, Darma Singh. (2003). *Food as medicine*. New York: Atria Books.

Kushi, Michio, with Stephen Blauer. (2004). *The Macrobiotic way: The complete macrobiotic lifestyle book*, Third edition. New York: Avery.

Porter, Jessica. (2004). *The hip chick's guide to macrobiotics*. New York: Avery.

Powers, Maggie. (2003). *American Dietetic Association guide to eating right when you have diabetes*. New York: John Wiley and Sons.

Smolin, Lori A. and Grosvenor, Mary B. (2005). *Basic nutrition*. Philadelphia: Chelsea House.

Webb, Robyn and Ard, Jamy D. (editors). (2004). *Eat to beat high blood pressure*. Pleasantville, NY: Reader's Digest Association.

# Index

Page numbers followed by the letter "f" indicate figures.

*Internet Guide to Medical Diets and Nutrition*
© 2006 by The Haworth Press, Inc. All rights reserved.
doi:10.1300/5852_16